Mobile Devices

Mobile Devices

TOOLS AND TECHNOLOGIES

EDITED BY **LAUREN COLLINS • SCOTT R. ELLIS**

CRC Press
Taylor & Francis Group
Boca Raton London New York

CRC Press is an imprint of the
Taylor & Francis Group, an **informa** business

A CHAPMAN & HALL BOOK

*In memory of Mr., my first love. Your
presence is felt in everything I do.*

Contents

Foreword

With the trend toward a highly mobile workforce, the acquisition of mobile device tools and technologies, such as the recent introduction of the Apple® iPhone 5S fingerprint scanner, handheld devices such as personal digital assistants (PDAs), and PC tablets, is growing at an increasingly rapid rate. These mobile device tools and technologies offer productivity in a compact form and are quickly becoming a necessity in today's business environment. Many manufacturers make mobile device tools and technologies using a broad range of hardware and software. Mobile devices are characterized by small physical size, limited storage and processing power, a restricted stylus-oriented user interface, and the means for synchronizing data with a more capable notebook or desktop computer. Typically, these devices have the capability to communicate wirelessly over limited distances to other devices using infrared or radio signals. Many handheld devices can now send and receive electronic mail and access the Internet. They also have built-in integrated seamless biometric security, like fingerprint scanners, to replace the old-fashioned passwords and passcodes.

Author Lauren Collins and the other contributors to this book have conducted both extensive research and practical hands-on development to promote, measure, and deploy emerging mobile device technologies, tools, and standards that revolutionize how devices are operated and used, with special emphasis on mobile systems and communications. The authors of this informative book make recommendations that utilize both analytical and empirical approaches, the development of simulation models, the buildup of proof of concept prototypes to evaluate new mobile device tools and technologies, and refined standard specifications for networks and systems. They also make recommendations that will help develop new test and measurement tools and metrics for next-generation mobile systems. The recommendations in this book encompass public safety communications systems, mobile ad hoc networks, body area network (BAN) standards, and secure and seamless mobility management protocols.

Furthermore, the authors make recommendations to improve the quality, reliability, resilience, robustness, manageability, security, and interoperability of mobile networked systems. As part of this nation's ongoing advanced mobile device tools and technologies critical national infrastructure program, the authors show readers how to conduct research and provide test, measurement, quality assurance techniques, tools, models, and reference data for emerging mobile device technologies, as well as how to develop, demonstrate, and promote these technologies through reference implementations, test beds, guidelines, and standards.

John R. Vacca
Author and editor of 76 books, including the best-selling
Computer and Information Security Handbook.
Pomeroy, Ohio

Acknowledgments

To my parents and grandparents for always supporting me and my relentless drive and will as I constantly strive to do and be more. I am forever grateful to all of you for everything you have done for me. Thank you.

My boss/mentor, Joyce. After changing my career and coming to work for the family company, she allowed me the time necessary to complete this project.

Amy Beth Outland, I could not have submitted this manuscript without your timely editing assistance. I know your manuscript, "I'm Not Broken," will be a best seller! Thank you for your knowledge and the family discount.

For Kayleigh, thank you for contributing your Instagram knowledge and artwork. Your social application skills make me nervous, but I am proud of you.

To my other half, thank you for tolerating me through countless hours in the middle of so many nights while I researched, wrote, revised, and completed this endeavor. Thank you for the endless supply of Red Bull, Mountain Dew, Propel, and Reese's Peanut Butter Cups.

John Vacca, an inspirational colleague and friend. Thank you for allowing me to be a part of your prior and future projects, and for referring this opportunity my way. To my contributors, thank you for your effort and for being a part of this special project. Rick Adams, Laurie Schlags, Tara Nieuwesteeg, and the Taylor & Francis team, thank you for your patience and for being available for my analytical questions as the project closed. All of you were truly amazing to work with.

Contributors

Lauren Collins
Founder and Chief Strategy Officer
Winning Edge Communications
Peotone, Illinois

Charlie Connor
Product Manager
kCura Corporation
Chicago, Illinois

Scott R. Ellis
Infrastructure Architect
kCura Corporation
Chicago, Illinois

Jonian Musa
Infrastructure Engineer
kCura Corporation
Chicago, Illinois

Nate Noonen
Application Architect
kCura Corporation
Chicago, Illinois

Josiah Roloff
Forensics Analyst
GCS Forensics
Chicago, Illinois

C.J. Wiemer
Application Specialist
kCura Corporation
Chicago, Illinois

Introduction

Mobile devices were designed to be portable handheld computers, to run a variety of applications, and oftentimes, to fit in your pocket. And they can serve the additional purpose of providing phone service on smartphones. While desktop and laptop computers are still necessary in order to use certain programs or perform certain tasks that mobile devices are not capable of, mobile devices provide an interactive experience for the user, and are designed primarily for consuming media. Some mobile devices are only able to connect to the Internet via Wi-Fi, while others have a data plan that allow the user voice and data service. Whether these devices are used solely for personal or business purposes, this technology allows us to work smarter and faster, perform tasks more efficiently with greater accuracy, and enhance communication and collaboration.

Mobile technology has enabled the continuous development of ubiquitous applications, providing immediate access to applications and data. Not surprisingly, users are accustomed to these powerful mobile devices, smartphones and tablets, which embrace virtually all aspects of communication while offering mobility and ubiquity like we have never seen before. In this book, we describe the fundamental concepts that comprise a mobile device; the software and application tools that make it run; the solutions that allow businesses to both build and provide better services, software, and solutions to their customers and providers; and the mobile services that are changing society as we know it.

Multiple books offer knowledge and instruction to build applications on one platform or another. Other books cover the end-to-end network infrastructure of cellular, wireless, and mobile communications, and some books simply inform the reader on system and device security. Our approaches to all of the topics outlined previously provide a high-level overview of systems, applications, networks, and services. This book is written for multiple levels of technological experience. Therefore, readers at any and all levels of technical expertise will fully understand the tools and technologies available to mobile devices.

Subsequently, this book is divided into four diverse sections:

Section I: "Mobile in Action." The first set of chapters provides basic knowledge regarding the use of mobile devices for both personal and business use. Chapter 1 educates readers on wearable devices and their features. We also discuss the tools remote workers can choose from, including cost, functionality, and feature sets of all named products.

Chapter 2 describes the rendering aspects developers must consider when developing mobile or web applications. We briefly cover the necessary considerations when developing for location tracking, retail, social media, and online purchasing.

Chapter 3 reviews the platform tools, preparing the reader for developing applications throughout the rest of the book; the tools and platforms used set the foundation for the size, growth, and purpose of the app.

Chapter 4 covers the angles of enterprise mobile strategies, the journey to develop and deploy a one-size-fits-all design, and how large organizations are purchasing components to stay on top of technology and ensure their organization properly forecasts consumer and business needs.

Section II: "Designing Mobile Apps." Chapters 5 to 8 focus on the entire application development process, from strategy and process through the development phases, to production and app launch. We level the playing field and cover both Android and iOS operating system development. Chapter 5 introduces the reader to the initial piece of app development: Consider the user, platform, and uses of the application.

Chapter 6 provides an in-depth description of push messaging. One major feature of app building is the ability to send push notifications to the users. We cover the architecture and process behind iOS push messaging in this chapter.

Chapter 7 discusses mobile databases and the API. Protocols and tools are covered at the developer level, considering the major features of the app as a whole, including monetization and value. Therefore, the framework and design must coincide with the business purposes as well as the user experience.

Chapter 8 is devoted to using the Agile software development methodology on the iPhone platform. This chapter covers the entire app building process and defines the processes development teams encounter from beginning to end.

Section III: "Mobile Services." Chapters 9 and 10 explore mobile access standards, as well as the services that cellular and wireless devices utilize.

Chapters 11 and 12 elaborate on the networking and infrastructure components that provide all the systems, access, and functionality leading up to this section. Although this section focuses heavily on cellular, wireless, and switching mechanisms, it is the foundation for the technology and tools utilized by mobile devices.

Chapter 13 discusses the different means of wireless communication and technology used by and designed for mobile devices. We describe the use of wireless telemetry in the medical field, the deployment of sensor networking capable of monitoring environmental conditions, and focus on state-of-the-art products designed for the consumer.

Section IV: "Security and Data Analysis." Chapter 14 provides an overview for awareness of mobile device threats, risks, and a few of the options on the market to counter the vulnerabilities.

Chapter 15 covers the use of digital forensics relating to recovery of digital evidence or data from mobile devices.

Chapter 16 discusses the various levels of security and types of encryption. Then we introduce types of attacks commonly seen in the industry.

Chapters 17 and 18 are devoted to mobile device management and corporate security. We also introduce the tools and applications corporations may implement for mobility and workforce management.

Depending on the readers' level of technical expertise, our discussions can be read easily from cover to cover or utilized as reference material. This book is designed to be flexible and enable the reader to move between chapters and sections to cover the material with ease. This book is written for readers at all levels of technical expertise, from developers to network engineers, to students preparing for a career in mobile technology or information technology managers looking to further explore and understand the value of mobile tools and technology.

I

Mobile in Action

CHAPTER 1

The Evolution of Mobile

Lauren Collins

CONTENTS

THE MOBILE EXPERIENCE has our undivided attention. Everywhere you go, whether commuting to work, out for a walk, on vacation, or while attending a sporting event, you see mobile technology in action. The relationship we have with our mobile devices is unlike any other—and it will continue to evolve as each generation leaves its mark. "It's not just about being connected, it's about being connected with a purpose," says Larry Irving.* Mobile devices provide users with voice, data, and multimedia service at any given time, place, and in any format. A mobile device is considered to be a handheld computing device, characteristically equipped

---

* Matt Petronizo, "How to Use Mobile Devices to Solve Global Problems," Social Good Summit, September 23, 2012, Mashable.com (accessed May 26, 2014).

with a touch screen. There are multiple mobile manufacturers and operating systems (OSs) for mobile devices. These devices are capable of running application software, commonly referred to as apps. The basic concept of mobile technology allows the user to carry one ubiquitous technology device, providing low subscriber costs, on a personable, customizable interface. The options are endless; mobile devices offer educational content, the use of proprietary software to collaborate, and provide access to all the tools needed when working remotely. And you can even run an app to monitor your heart, which dynamically sends your vitals to your doctor. Familiarity with the benefits of mobile devices, and the ways to implement the synergy of mobile technology, and the limitless opportunities available is an essential skill to possess.

The adoption of mobile devices has been an exciting and overwhelming advancement to both business and the personal lives of everyone, and the trend continues to grow. As such, the model of mobile technology presents opportunity: a wide scope of deployments, innovative technology, and bleeding-edge tools must be provided to the user. These technologies include the design, integration, and support to embrace usability, communications, scalability, and mobility.

Mobile technology hardly refers to voice and cellular, but now focuses instead on the types of technology available to consumers and businesses. What had started as a phenomenon to solely keep in touch and be reached by phone has evolved into an essential tool in both our personal and our business existence. Even though apps seem to have been the hot item when marketing mobile devices, app usage is second to browser usage on mobile devices; however, app access and web browsing are not mutually exclusive. Figure 1.1 indicates that mobile users have quickly come to utilize technology and enjoy the experience of a mobile device, working to create an interactive experience.

Mobile technology was forced to evolve from cellular technology, relative to rising above both performance and scalability limitations. Cellular carriers recognized an exponentially vibrant growth in the subscriber base as well as the desired usage per demographic. Carriers could no longer offer unlimited data to their customers and continue operating under the same service model to support more traffic and users. Continued growth of bandwidth oversubscribes the link, and mobile packets at the core layer of the network suffer. Carriers have implemented additional infrastructure and bandwidth to support the capacity and needs of subscribers, but this engineering is ongoing. Mobility is a crucial consideration

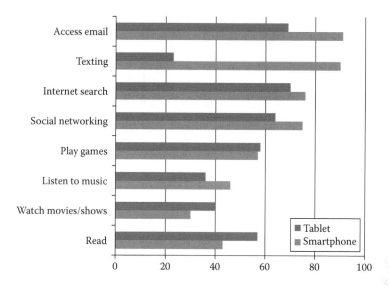

FIGURE 1.1   Daily activities on a smartphone and tablet performed at least once a day.

providers must recognize; the industry has shifted its perspective of cellular, or wireless, to a service and device-centric mobile model.

1.1 INTEGRATION OF MOBILE TECHNOLOGY

The average individual now possesses a minimum of two smart devices. Children of only a few years of age crave the use of a tablet or mobile device. Wireless and mobile communications have grown exponentially, and will continue to grow as the need for technology has embedded itself in our lives. Today, most consider the idea unthinkable to exist without constant and continuous communication and connectivity to the Internet. The evolution and growth of mobile technology has fundamentally altered the use of the Internet, including the means by which content is accessed on the Internet, utilizing both smartphones and devices such as tablets and netbooks.

The consistent advancement of mobile devices feeds the ever-growing appetite for faster bandwidth, uninterrupted connectivity, applications to fulfill the needs of consumers and businesses, and the security of all of this information. The number of users who access the Internet on a mobile device will soon surpass the number accessing it on a PC. Table 1.1 lays out the activities users perform on mobile devices and laptops and workstations and portrays the interest shifting to mobile.

TABLE 1.1 Consumers' Desired Behaviors and Preferences of Mobile Devices vs. Desktops and Laptops

Category	Prefer Mobile Device	Prefer Laptop/Desktop
Social networking	60%	40%
News and information	35%	65%
Music	70%	30%
Play games	60%	40%
Watch TV/video	65%	35%
Sports	40%	60%
Maps/directions	65%	35%
Travel arrangements	35%	65%
Check travel status	80%	20%
View account balance	75%	25%
Conduct bank transactions	45%	55%
Check weather	50%	50%
Shopping	35%	65%
Email	55%	45%
Work activities	20%	80%

Smartphones and tablets outsell computers by four times, and many will have their first experience with the Internet on a mobile device rather than a computer. In 2013, more than 45% of network devices were shipping without a wired port. The need is now greater than ever to ensure wireless speed is as fast and reliable as wired media. Mobile devices are energy-efficient, portable, and command less processing power and a condensed ability to run resource-intensive operating systems. These devices should be understood as an interface to the Internet rather than as a computing device. Mobile technology is geared toward the user who will access cloud applications, email, and general purpose apps.

Consumers have resorted to checking balances on mobile devices rather than driving to an ATM and printing a paper balance receipt. Mobile devices may be utilized in personal or professional instances, deeming either equally useful. Rich interfaces and supported user interaction enable collaboration and a more immediate response with the application. These interactions include swiping, pinching, and rotating, to name a few. Speech recognition is another key capability of a mobile device. Whether a traveler uses an app for language translation or to search for directions, speech analytics open up another world pertinent to consumer intelligence and marketplace strategies.

1.2 WEARABLE TECHNOLOGY

When carrying multiple items, how many times were you unable to answer an important call, let alone know who the caller was? Wearable technology delivers many useful tools for people with active lifestyles. And it's all possible, thanks to technology. Our culture and lifestyles usually shift daily, to accommodate technology. The question is no longer "What do we do with technology?" But instead it is "How will we manage technology rather than it managing us?" Those who are embracing technology have been part of the shift in the economy, the way business is handled, and the shape of our entrepreneurial future.

The mobile market has successfully reeled an entire economy into using a mobile device for their end-all, be-all source of knowledge. Whether you need to know the time, date, or directions to a specific location, your mobile device is likely to be your preferred method to access most information. The purpose of wearable technology is to integrate itself into an adjacent sector of the mobile marketplace. Active people have the capacity to track their workout activity. This allows users to dynamically sync to an app, as well as listen to music from their cloud account on their smartwatch. Manufacturers and vendors are anticipating wearable technology to provide more convenience, more positive user experiences, and more comfort than carrying their traditional mobile phone or tablet around with them.

1.2.1 Google Glass

Google® Glass™ is the most popular and well-known brand of wearable technology today. Everyone has heard of it, but how many have tried it? The potential is sensational. This wearable device simulates a portion of the communication means from your mobile phone or tablet and shows it to you in a heads-up display. You can view, respond to, or delete emails as well. Google Glass also has the ability to take calls, take pictures or videos, and is aware of your location. So, whether you need a map or you would like Google Glass to guide you to your destination, it can display the information you need.

Google Glass runs on the devices. It connects via Wi-Fi or Bluetooth to a phone, a laptop, or a consumer-grade Wi-Fi network. The frames are titanium and therefore are much lighter than you would think. Pictured in Figure 1.2, Glass has a thick, transparent display that sits above your

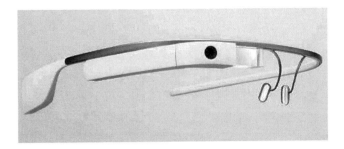

FIGURE 1.2   Google Glass, one of the most anticipated forms of wearable technology, is pictured here. Unfortunately, Glass has not been released to the public yet and is still undergoing beta testing.

right eye, and the right arm wraps around, which encompasses a mobile computing device that is built in to the stem that sits on your right ear. There is a battery housed on the right side as well, where the buttons, speaker, and touch pads are located. Currently the battery only lasts up to a day, and more commonly between 1 to 2 hours with normal use. The iteration that is out at press time does not have a volume control, making it impossible to hear phone calls in public areas.

While some may wonder if this wearable device will soon replace our smartphones or tablets, there are several irritating limitations presented by this beta product. For instance, Google Circles must be set up using a web application prior to sharing any photos or videos. In addition, at this time Google Glass users can only add up to 10 contacts, but the contacts are restricted to your Gmail™ contact list. The only email supported is Gmail, so no corporate integrations for Outlook or support for other email types. And if you do not have an Internet connection, its only use is as an offline camera.

Google Glass is controlled by voice and swiping and tapping gestures. The functionality is basic at this time, and a great deal of the management of the device takes place on the Internet and not Glass itself. For instance, if you were connecting to a Wi-Fi network, the best route is to download the MyGlass app to add networks. The alternate method is through the MyGlass site in a web browser, where you enter each network's service set identifier (SSID) and password. A QR Code® is then generated that you look at with Glass, which will then log in and remember the network. For anyone who is curious and has not seen the interface yet, it is very similar to Google Now™.

FIGURE 1.3 TrackingPoint's Precision Guided Firearm (PGF) utilizes wearable technology.

1.2.1.1 Precision Guided Firearm

Even the military may soon benefit from an app that pairs wirelessly with wearables, such as Google Glass. TrackingPoint™, a company who develops precision firearm tracking technology for firearms, has built an app called ShotView.* The Precision Guided Firearm (PGF) system uses optics and fire control, enabling a shooter to be more accurate at longer ranges than he or she is capable of. TrackingPoint's Tag Tract Xact™ system technology enables the user to tag a target by pointing and clicking, then dynamically setting up the shot, incessantly tracking a static or a moving target. There are a handful of ballistic variables, including target velocity, range, wind, temperature, pressure, and shot angle. Rifle cant is instantly calculated into an optical firing solution. This app streams video from the firearm's tracking scope to mobile devices such as wearables, phones, or tablets. The shooter will realign the reticle with the tag point and pull the trigger. Since the PGF's guided trigger is connected to tracking optics, say the shooter was misaligned, the PGF-guided trigger will increase in weight and push back on the shooter to defer firing until the shooter realigns. Figure 1.3 illustrates the fascinating components to bring this technology to a wearable device.

1.2.2 Smartwatches

Smartwatches are wearable devices that operate a multitude of tasks, comparable to wearing a tablet. Mobile apps run on some smartwatches, and mobile OSs run on the remaining models. Certain models even support phone calls, while others display text messaging or the subject line of new emails.

---

* "Google Glass Gun App Would Provide 'Mind-Blowing' Ability to Shoot around Corners," RT, 40, January 2014, RT.com/usa (accessed June 7, 2014).

1.2.2.1 Pebble

The Pebble® smartwatch, pictured in Figure 1.4, stands out above all others simply for its ability to have firmware upgrades. Pebble is available in both Android™ and iOS platforms, and its most recent firmware update to version 2.2 added iBeacon™ technologies by Apple®. iBeacon utilizes Bluetooth low energy proximity sensing to transmit a unique identifier. This identifier has the ability to identify the location of a device using an Internet connection, then transmit the information or prompt an action such as a push notification.

This firmware update added the following feature sets in the SDK 2 (software development kit):

- Ability to call web services through the phone
- Persistent storage offering
- Supports GPS information from the phone
- Accelerometer
- Added iOS and Android portable mobile framework
- Dynamic memory and memory protection
- Data logging

FIGURE 1.4   The Pebble smartwatch works with both Android and iOS platforms.

Pebble also integrates with the GoPro camera, which is valuable for users who are recording an intense activity.

1.2.2.2 MARTIAN

The MARTIAN™ does not resemble the typical smartwatch. It is a sophisticated version of a watch that works in conjunction with a mobile phone. The MARTIAN presents hands-free voice commands, messages and alerts display, and smartphone camera control. This smartwatch is a perfect smart device for an executive who appreciates technology. Figure 1.5 displays the two options available, the Voice Command, having both notifications and voice command capabilities, and the Notifier, only providing notifications. Custom vibration patterns can be configured for alerts and notifications in an effort to distinguish who is trying to reach you.

MARTIAN recommends using iPhone 4S and later for use with Siri®, and Android 4.1 Jelly Bean or later for use with Google Now. Once the watch is powered on, your phone will pair with the smartwatch using Bluetooth. The MARTIAN Alerts app is where the user can set up notifications and toggle settings. Android allows you to choose which apps you receive alerts from; however, the iPhone *only* supports calls, text messages, calendar alerts, reminders, email, and Facebook and Twitter messages. And the biggest downfall with the iPhone is that it will not send you push email, and can only use internet message access protocol

FIGURE 1.5   Pictured left to right: MARTIAN Notifier allows the user to program customized vibration patterns for different types of notifications; MARTIAN alternate style.

FIGURE 1.6 The micro-USB connection is located on the side of the MARTIAN and is used to charge the smartwatch as well as to reset the software if any technical issues occur.

(IMAP) or post office protocol (POP) accounts in the MARTIAN Alerts app. Only one email account is supported, which is hardly the use case for anyone today. Charging is supported using a micro-USB port, shown in Figure 1.6.

1.2.2.3 Neptune Pine (in Development)

The Neptune Pine® is an Android smartphone that you wear on your wrist. Unfortunately, this product is still in development, but you can preorder it on Android's website. The Pine has a 1.2 Ghz dual-core SnapDragon processor, 32 GB of memory, up to 64 GB of storage, a 2.4-inch full-color Gorilla Glass display, a front-facing camera, and a 5-megapixel, rear-facing camera. The functionality and specifications contained in this device could very likely replace a smartphone. The Pine utilizes a QWERTY keyboard, shown in Figure 1.7, and is as easy as typing on a smartphone.

FIGURE 1.7 Neptune Pine employs a QWERTY keyboard, making typing similar to using a smartphone.

FIGURE 1.8 The Hyetis Crossbow offers every smartphone software feature you can think of, plus several other smartwatch offers, such as facial and gesture recognition.

1.2.2.4 Hyetis Crossbow

The highlight of the Crossbow® certainly is the 48-megapixel camera with Zeiss optical zoom lens and integrated flash. The product is one of a kind in this market, boasting a MARSO gesture and face recognition device. Device communication with the Crossbow can be done over Wi-Fi, Bluetooth 4.0, or near-field communication (NFC). Environmental data can be delivered to your smartphone with a variety of sensors: temperature, altitude, depth, and GPS, to name a few. The depth rate for scuba diving is 15,000 meters. The Crossbow is shown in Figure 1.8.

1.3 BUSINESSES GO MOBILE

Business professionals would choose to give up their morning coffee over their smartphone. Mobile devices are playing a progressively essential role in the methods people use to get work done and stay connected. Before, phone, email, and calendar access were the only uses for mobile devices in the workplace. Now hundreds of apps have evolved to meet the needs of enterprise business users. The ability to be mobile has empowered both employees and customers. Companies who are adaptable and can welcome the continual increase of mobile technology are going to succeed.

Executives and business decision makers have realized that mobile devices can ultimately help them achieve their objectives and work from

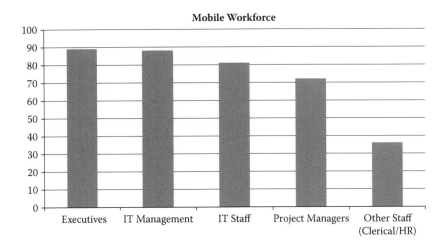

FIGURE 1.9   Firm leadership and IT are power users on mobile devices for work-related activities. Other professionals have come up with very productive ways to take advantage of mobile devices, and these will continue to grow over the years.

any location. Checking emails, downloading files, and accessing the web are now easier and faster thanks to the power and speed of the 4G and LTE (Long Term Evolution) networks. With the requirement to work from any location comes the need for a secure and seamless connection that does not require the use of a traditional virtual private network (VPN). As our businesses implement mobile technology, employees are not able to do their jobs effectively without using their mobile devices. While most of the mobile workforce is made up of executives and IT departments, Figure 1.9 shows all levels that are using mobile devices for work-related activities.

As mobile devices become more sophisticated and apps are tailored to fit business needs, we will continue to see a growing need for more and more mobile devices in businesses. Smartphones, tablets, and other mobile technologies are driving productivity in business, and in some cases introducing innovation to firms who have fought the power of technology for decades. As a result, anyone who has a website or runs an online business must consider the mobile experience. A business used to build a website when it opened its doors. Today businesses have either deployed or are developing a mobile-friendly website or mobile app. Figure 1.10 illustrates the important need for businesses to deploy at least one of the mobile solutions for their customers or clients. If most people find your business through an Internet search, then optimizing your site for mobile web should be a high priority. However, if an immersive experience attracts

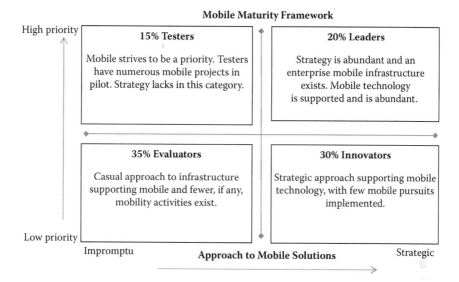

Mobile Maturity Framework

| | |
| High priority |
15% Testers	**20% Leaders**
Mobile strives to be a priority. Testers have numerous mobile projects in pilot. Strategy lacks in this category.	Strategy is abundant and an enterprise mobile infrastructure exists. Mobile technology is supported and is abundant.
35% Evaluators	**30% Innovators**
Casual approach to infrastructure supporting mobile and fewer, if any, mobility activities exist.	Strategic approach supporting mobile technology, with few mobile pursuits implemented.
Low priority	
Impromptu	**Approach to Mobile Solutions**	Strategic

FIGURE 1.10 An organization's Mobile Maturity Framework is centered on two components: the extent to which mobile is regarded as a priority and the degree to which it is treated as a deliberate ambition. Four phases of maturity are identified.

your clients, package the mobile site and app together with the same order of importance.

1.3.1 Defining a Mobile Strategy

With the continuous rise of mobile use in organizations, mobile device management (MDM) and mobile application management (MAM) continue to be high on the list of priorities for IT managers. Throughout this book, the authors will touch on the multiple flavors of devices and how each device has individual characteristics and intricacies, which can be the difference between opening up your organization to hackers and being protected. Figure 1.10 illustrates the relationship between firm priorities and their approach to implementing a mobile solution. Two measurements quantify the Mobile Maturity Framework: the priority for mobile technology projects and the magnitude to which mobile is treated as a strategic initiative. Regardless of the urgent need to focus on mobile use in the enterprise, only 20% of technology leaders are prioritizing and implementing a mobile strategy.[*]

_____

[*] Stephen D. Drake and Rebecca Segal, "Putting Mobile First: Best Practices of Mobile Technology Leaders," IBM Global Technology Services, Somers, NY, June 2013.

We can identify four phases of mobile maturity within organizations. There are *evaluators*, which can be categorized at the earliest stage of mobile adoption. Evaluators do not have a strategy defined for the business, let alone at the business unit level. They have an impromptu methodology and may not have any mobile technology initiatives in the hopper. *Testers* identify mobile as a priority and have a number of mobility projects underway; however, testers have an approach that implements mobile at the business unit level, not as an organization. *Innovators* evaluate or roll out a limited number of mobile initiatives. They may enforce policies for mobile devices, but have a conventional methodology to a full deployment of mobile technology until the business defines a solid return on investment (ROI). *Leaders* are the final phase within the Mobile Maturity Framework. Leaders are organizations that are able to identify an enterprise-wide, strategic approach to mobile technology and already have an infrastructure implemented. They support mobile technologies and have a large number of mobile-oriented initiatives incorporated across businesses.

Leaders of mobile technology:

- *Govern* mobile deployments for efficiency and security. Policy management is implemented by device and by application.

- *Optimize* mobile infrastructure for reliability and accessibility. The network requires constant monitoring and tuning to support increased access, storage, workload, and velocity created by the transition to mobile.

- *Integrate* mobile throughout the enterprise. Leaders correlate significant applications between business units, rather than treating mobile as a one-off project or silo project. This integration supports monetization of mobile initiatives and channel prospects.

- Make *strategy* for mobile development a top priority. Business cases for mobile are defined and roadmaps are developed based on those use cases.

Technology leaders shifting to mobile have experienced an increase in interactions with customers, an increase in profits, and more efficient employees, as shown in Figure 1.11.

The integration of mobile can have a significant impact on the financial success or failure of a business. Mobile technology must be integrated

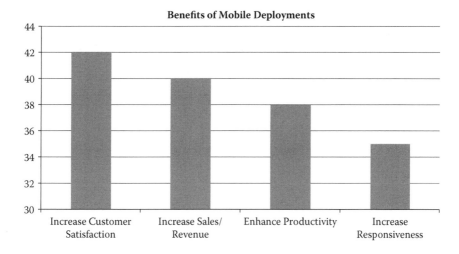

FIGURE 1.11   Mobile deployments are utilized as a productivity tool to a system of engagement.

into the business's roadmap, and the business strategy and infrastructure must align to meet consumer needs, while simultaneously producing profits. Organizations that implement a mobile strategy and concentrate on mobile technology initiatives are going to experience revenue growth and increased employee efficiency across the board. If a company supports the business unit initiatives, whether it is sales and marketing, customer service, or information technology, the quality and level of interaction will improve tremendously.

1.3.2 Working Remotely

Stellar customer service is oftentimes the foundation of a thriving business. By implementing mobile technologies, organizations have found their employees are more engaged when using tools that deliver a rich interface and provide timely and relevant information to communicate with coworkers and clients. Transactions are completed much faster with the integration of mobile technology, not to mention with more accuracy and attention to detail. Providing access to key functionality components drives enterprise efficiency, enabling employees and customers to interact and communicate with each other from any place, at any hour, utilizing the device of their choice. With the convenience of a mobile device and proven increase in productivity, many businesses are allowing employees to work remotely.

There are numerous remote access services and apps available to employees, but businesses must consider the implications for enterprise IT when it comes to security and compliance. For many organizations, the risk of a rogue app is a top concern of an IT security administrator. Multiple apps are blocked, which can cause frustration and a loss of productivity when a person is working from home. Most likely your IT department has defined one or multiple apps that have enterprise capabilities, such as password authentication, data encryption, or an app-specific micro-VPN. In addition, and just as important, support staff must provide comparable features and compatibility across multiple device platforms. The mobile device operating systems are interchangeable with most app developers. Whether your preference is iOS®, Android, or Windows® operating system, your organization will likely support the platform.

The most common remote access apps provide similar features and are available for all device types, and usually will possess the ability for an employee to access a Windows-based or Mac® desktop or server. Select your remote access app carefully, as it is always a great idea to read the fine print.

1.3.2.1 Remote Access Applications

1.3.2.1.1 PocketCloud   Wyse Technology® developed a fairly straightforward app that has both a free app for Android and iOS devices and a reasonably priced Pro version. There are three connection options: Remote Desktop Protocol (RDP), Virtual Network Computing (VNC), and Auto Discovery (shown in Figure 1.12). The RDP connection provides connectivity, but is not as intuitive as the VNC connection option. The VNC option allows snapshots to be taken and sent to the user for collaboration. The Auto Discovery option is linked with a user's Google® account and utilizes Google services to locate devices available to that user. The Pro version supports 256-bit encryption for RDP connections, compared to 128-bit encryption for the free app. The Pro version also supports third-party app integration, VMware View virtual desktops, and can interact with Microsoft® RD Gateway. There is a Premium version of PocketCloud for iOS devices that allows for file, picture, and video uploads from them to desktops, and also supports downloads, printing or emailing of files, and video streaming over speeds as low as 3 G.

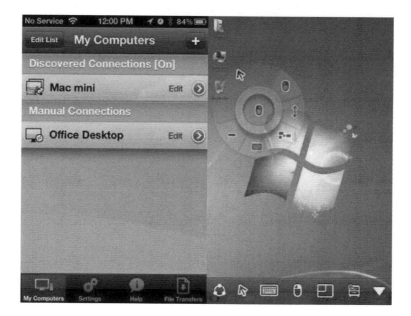

FIGURE 1.12 PocketCloud has three connectivity options. Auto Discovery and Remote Desktop Protocol (RDP) are pictured here.

1.3.2.1.2 GoToMyPC   Citrix® developed GoToMyPC® in a free version for its mobile app, but charges a fee to use the service. Three service plans options are available: corporate, pro, and individual, and have flat pricing either monthly or per year. All three plans have similar capabilities, but the corporate plan presents user management, group management, and the capacity to configure security and settings to determine which features employees or groups can use. GoToMyPC protects data with end-to-end 128-bit encryption and supports dual-password authentication, whereas each device is assigned its own passcode. Users have the ability to grant temporary users sharing permission or full control of their desktops. GoToMyPC does not offer the ability to move files between mobile devices and desktops. Figure 1.13 displays the virtual mouse navigation techniques.

1.3.2.1.3 LogMeIn   LogMeIn® offers two plans, free and pro. Its free version supports remote access for up to 10 devices, clipboard syncing, keyboard mapping, multiple monitors, and wake-on-LAN (allows a network message to be sent to turn on or wake a device). The Pro version, formerly

FIGURE 1.13   Mouse navigation techniques GoToMyPC utilizes on mobile devices.

LogMeIn Ignition, has additional features, such as file and desktop sharing, video and audio streaming, remote printing, and cloud services. The pricing structure for LogMeIn can be very unclear. There is a mixture of conflicting prices at any given time, depending on promotions. LogMeIn is displayed in Figure 1.14.

FIGURE 1.14   LogMeIn offers cloud services integration and file sharing on the Pro version.

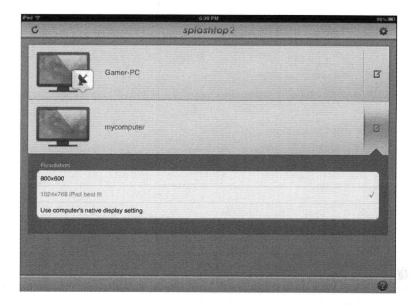

FIGURE 1.15   The recent release of Splashtop2 provides added security for the infrastructure that allows users to connect to their devices in the cloud.

1.3.2.1.4 Splashtop   Splashtop® allows users to view and edit files, access applications, stream videos, and play graphic-intensive games. There are three levels of service offered: enterprise, business, and personal. On top of the levels of service, there are feature packs that can be added on to the business and personal plans. One pack supports 256-bit encryption, and the other pack supports a whiteboard service for annotating a live remote desktop screen. Business users are provided with centralized administration, group management, and the ability to maintain and monitor an audit trail. Splashtop's Enterprise version supports hosted apps, virtual desktop access, Active Directory integration, and the ability to integrate with third-party mobile device management tools. Splashtop2 is pictured in Figure 1.15.

1.3.2.1.5 TeamViewer   TeamViewer® offers a sophisticated product that is geared toward tech support and online teamwork. The product is expensive, but provides the ability to facilitate online meetings or support other devices. TeamViewer uses 256-bit encryption and utilizes a proprietary protocol to ensure secure connectivity. TeamViewer QuickStart is a free desktop application that is designed to work with the full version to support

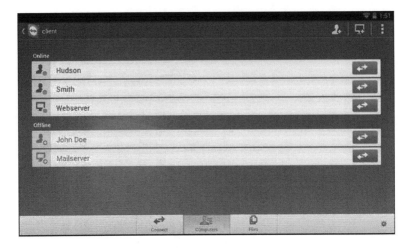

FIGURE 1.16   TeamViewer offers a great package for businesses to support remote devices and also allows for team collaboration.

multiple client devices. Figure 1.16 displays the view for saved devices frequently connected to it, whether they are online or offline, allowing for easy connection to those devices. So, if you are supporting mobile devices, this is a great product in the market.

1.4 SUMMARY

Mobile devices comprise the largest consumer electronics market in the world. Therefore, mobile devices are subject to cutting-edge innovation in multiple different areas. In order to remain competitive, devices, carriers, OS, and software vendors must stay up to date and in line with this rapidly changing market. Mobile technology places an emphasis on connectivity, displayed content, user experience (UX), gesture, speech, and facial recognition. Security, infrastructure, and user interface (UI) are also areas of high demand. All of us have and will continue to encounter mobile technology, and understanding the technology trends allows us to be aware of what lies ahead.

Site Detection

Lauren Collins

CONTENTS

I N THE MOBILE WORLD, knowing a user's location permits an application to be smarter and deliver enhanced information to that user. What do you do on your mobile device most of the day? Be honest—the majority of our time spent on mobile devices in a given day includes obsessing over information on your mobile device. Whether mapping out the evening activities based on reviews and ratings, making reservations, or literally viewing a map to navigate between dinner and a club, we basically cannot make decisions anymore without consulting our mobile devices. This chapter outlines the considerations developers and users must make regarding the countless types of devices available, the platforms and application program interfaces (APIs) to contemplate working with, and the way content display in its various forms can either enhance or annihilate the user experience.

We choose our mobile devices based on the functionality and features that meet our distinct needs. A mobile device, small in size and easy to take everywhere, is ubiquitous and rich in its uses. While working or attending a sporting event, your mobile device camera allows one to capture and post pictures and videos, add comments or view others' comments, and literally stay in touch 24 hours a day—even on holidays. Individuals check their social networks and email throughout the night, waking from sound sleep, and in the morning upon waking up. If this sounds like you or someone you know, you are not alone. This chapter is dedicated to the social impact that mobile devices have had on our lives. And this impact centers largely around the desire for location-based information.

2.1 SHOPPING FOR INFORMATION

A large number of mobile device owners use their devices for shopping-related activities. Figure 2.1 demonstrates the behavior between smartphone and tablet users for shopping-related activities.

All sorts of retailers, both online and in store, are looking for information about shoppers to capture their habits, track their movements, and have technology in place to track customers' movements by following the Wi-Fi signals from their mobile devices. Physical store retailers use these tracking methods in an effort to decide on whether they should change store layouts, shelf stocking techniques, and offer customized coupons. There are device-aware stores and device-aware locations. Technology analytics make it possible to measure the distance between a smartphone and Wi-Fi antenna to count the number of patrons walking past a store and track how many actually enter. Online retailers have the ability to track cookies, which is not entirely different from the ways physical retailers

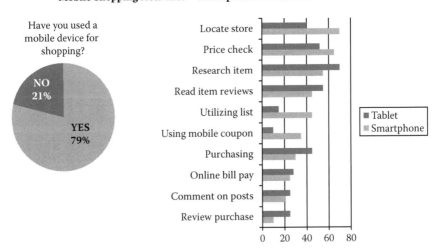

FIGURE 2.1 Smartphone and tablet users are similar in their behavior.

track behaviors. If a shopper's phone is set to look for Wi-Fi networks, a store that offers service can pinpoint a shopper's location in the store, within a 20-foot radius, even if the shopper does not connect to the network. The store can also recognize returning shoppers, because mobile devices and unique identification codes propagate when they search for networks. Now stores can tell how repeat customers behave as well as the average time between visits.

The cameras placed in stores are so sophisticated, with sharper lenses and data processing, that companies are able to analyze what shoppers are looking at, and even detect their moods. These cameras give analysts the ability to monitor facial cues at the register, as they walk throughout the store, or even gauge their response to online ads. Marketing companies now offer software for the cash register that tailors marketing messages to a customer's age, gender, and mood, all measured by facial recognition. Figure 2.2 shows an aerial camera view for a retailer to track customer movements, the amount of time spent in a particular area in a store, the amount of time spent in the store, and whether an item was purchased or not.

FIGURE 2.2 Retailers employ cameras to track consumer shopping patterns, including the length of time spent in the store, whether an item was purchased, and to gather data about the shopper's behavior.

When a shopper volunteers personal information, either by downloading a retailer's app or providing an email address when using in-store Wi-Fi, the store is then able to pull the customer profile, the number of visits, what products the customer was looking at on the Internet the past week, and his or her purchase history. Based on these types of analysis, a store may know from information gathered during the past week whether a shopper is interested in shoes or outerwear, and that store may send an email with a coupon specifically tailored to the item you have been searching for. While these methods may seem intrusive, some consumers are happy to trade privacy for deals. Certain companies offer apps to track your shopping activity in exchange for cash and prepaid gift cards. Anyone who agrees to this type of information tracking has voluntarily agreed to be tracked over GPS, Wi-Fi, and cellular networks. The data gathered can be sold to store owners, online retailers, and app developers.

There are four tiers of customer tracking:

1. Each time a card is swiped, a unique identifier is associated to it. Retailers identify trends involving things like repeat business, even though they may not know who each customer is.

2. When the cashier asks if the shopper would like a receipt emailed, the email enters the tracking system, and the store can now identify the cardholder. The store can send the customer targeted ads and coupons.

3. A software platform indexing information from 1 and 2 serves as the backend to a store's location-aware mobile app, which runs on customer phones. The app will push coupons and offers to the shopper's phone while he or she is in the store.

4. When asked "Can we send promotional emails?" the customer has just agreed to an opt-in service where physical retailers can push targeted advertising across web-based ad networks.

Many people have thought at one time or another, "How did the store know I have been looking at kayaks?" All of your online and in-store shopping is being tracked, and the retailer is looking for the perfect opportunity to reel you in. Customers feel like they are receiving personal attention as a result of retailers positively shaping the user experience. And if this seems creepy to you, let this serve as a reminder that there are real live people on the other side of every Google® search, Amazon® purchase, and Facebook®

like, and they are using what they know about you to make money. In addition, chapters in this book dig deeper into location-awareness technology, near-field communications, cellular triangulation, and global positioning system (GPS) functionality and how these technologies can generate so many capabilities for businesses and personal use of proprietary information.

2.2 DEVICE AWARENESS

Device awareness is the process of consuming information in a web request to identify a mobile browser or device and ascertain its capabilities. Learning the mobile device's characteristics allows a mobile website to make determinations about adapting mobile styles, scripting, and page layout in an effort to deliver an enhanced user experience. Content adaptation can use device characteristics, which are obtained through device awareness, as criteria for changing the functionality or design of the mobile site. Mobile device detection generally answers two questions: "What is it?" and "What can it do?" Once the device has been determined, proper site redirection may take place for the mobile device to properly display content rendering.

Many developers will work to ensure users are supported across all types of devices. There are many ways to approach this, and ultimately all of the following will work.

Do nothing: The easiest method to support mobile devices is to place a link somewhere near the top of the page that points to the mobile version. Users will be able to choose whether or not they want to see the optimized version or continue with the normal web version. The biggest downfall is that both versions must be updated, and over time and as the sites grow to be larger, this becomes a daunting task.

Use JavaScript: Most people try JavaScript first. Then they quickly realize this is the worst option. Even though browser detection scripts seem like a good idea, there are thousands of mobile device types out there. Attempting to detect the mobile device types with one JavaScript script will quickly become a catastrophe.

Use CSS @media handheld: Using the content services switches (CSS) command @media handheld is another way to display CSS styles just for mobile devices. The developer writes one web page, and then creates two separate style sheets. The first sheet is for the screen media types, which styles the page for monitor display using a computer. The other is for the mobile device; it styles the page for smartphones and tablets. If you have ever tried this, you already know that many devices do not support the *handheld* media type; they display their pages with the *screen* media type instead. Older devices do not support CSS at all.

Use PHP, JSP, and ASP to detect user-agent: It is possible to redirect mobile users to a mobile version of a site, because it doesn't rely on a scripting language or CSS that the mobile device does not use. Instead, it uses a server-side language (PHP, ASP, JSP, ColdFusion, etc.) to look at the user-agent and then change the HTTP request to point to a mobile page if it's a mobile device.

The issue is that there are tons of other possible user-agents that are used by mobile devices. This script may catch and redirect a large majority of them, but not all of them. Also, as with the other solutions, two separate sites would need to be maintained for the users.

Use WURFL: Wireless Universal Resource File (WURFL) is a decent solution to redirect mobile users to a separate site. WURFL is composed of an XML file, database (DB) file, and various Perl database interface (DBI) libraries that contain not only updated wireless user-agent data, but also the features and capabilities those user-agents support. To implement WURFL, download the XML config file, then choose the language and implement the API on your website. Currently, there are tools for using WURFL with Java, PHP, Perl, Python, Ruby, and C++.

The benefit to using WURFL is that there are numerous people updating and adding to the config all the time. So while the file you may be using is outdated almost before you finished downloading it, chances are you will have all the mobile browsers your readers routinely use without any problems. In addition, WURFL does more than simply detect the user-agent; it tells you which devices support what. For example, if you elect to set up a Wireless Application Protocol (WAP) push, you could provide that service only to the devices that support it. Other devices will not even see the link.

Use responsive design. The best solution for detecting mobile devices is to use responsive design on your existing pages. CSS media queries are used to define styles for devices of various screen widths. Responsive design allows you to create one web page for both mobile and nonmobile users. This way you do not have to worry about what content to display on the mobile site or remember to transfer the latest changes to the mobile site. Also, once you have the CSS written, you do not have to download anything new.

Responsive design does not usually work well on older devices and browsers, but because it is additive (adding styles onto the content, rather than taking content away), readers will still be able to read the website; it simply will not look ideal on an older device or browser.

Create separate versions for each device class: All the options listed above are great if you are a home user, but in reality none of them will work if there are any concerns with performance or serving different devices' classes. Divide your devices into separate classes and design the best experience for each category. Figure 2.3 displays the three common devices types: (1) small touch screens—smartphones, (2) large touch screens—tablets, and (3) large touch screens with keyboard and mice—laptop-tablet combination. eBook readers are not listed in the three categories since they usually have screen reader software installed and work fine if you build in accessibility.

FIGURE 2.3 Create a great user experience by categorizing devices into three classes: (1) small touch screens, (2) large touch screens, and (3) large touch screens with keyboard and mice.

2.2.1 Server-Side Detection

The server presents a much more limited understanding of the device you are dealing with. The most useful indication available is the user-agent string, which is supplied via the user-agent (UA) header on every request. As a result of this, the same UA sniffing approach will work here. Projects that already do UA sniffing are WURFL (open source) and DeviceAtlas (licensed), but they include an enormous amount of additional information about the devices. There are challenges to all the products out there. WURFL can grow to be very large, containing around 50 MB of XML, which theoretically incurs significant server-side overhead for each request. And there are several free projects out there that are not only less comprehensive, but will also only distinguish whether a device is mobile or not. Furthermore, those types of projects provide only limited tablet support through an ad hoc set of tweaks. Figure 2.4 illustrates an example of open-source mobile phone detection.[*]

If you do not have the development staff on hand but still want to provide a great mobile experience, there are a handful of vendors out there that make this process easier. ScientiaMobile® provides four products to support your specific need for control, scalability, and flexibility. Figure 2.5 describes the four products ScientiaMobile offers and their respective features and the functionality of each.[†]

```
def redirect_mobile(url = "http://detectcollinsmobilebrowser.com/mobile")  redirect_to
url if /(android|bb\d+|meego).+mobile|avantgo|bada\/|blackberry|blazer|compal|elaine|
fennec|hiptop|iemobile|ip(hone|od)|iris|kindle|lge |maemo|midp|mmp|mobile.+firefox|
netfront|opera m(ob|in)i|palm( os)?|phone|p(ixi|re)\/|plucker|pocket|psp|series(4|6)
0|symbian|treo|up\.(browser|link)|vodafone|wap|windows (ce|phone)|xda|xiino/i.match
(request.user_agent) || /1207|6310|6590|3gso|4thp|50[1-6]i|770s|802s|a wa|abac|ac(er|
oo|s\-)|ai(ko|rn)|al(av|ca|co)|amoi|an(ex|ny|yw)|aptu|ar(ch|go)|as(te|us)|attw|au(di|
\-m|r |s )|avan|be(ck|ll|nq)|bi(lb|rd)|bl(ac|az)|br(e|v)w|bumb|bw\-(n|u)|c55\/|capi|
ccwa|cdm\-|cell|chtm|cldc|cmd\-|co(mp|nd)|craw|da(it|ll|ng)|dbte|dc\-s|devi|dica|
dmob|do(c|p)o|ds(12|\-d)|el(49|ai)|em(12|ul)|er(ic|k0)|esl8|ez([4-7]0|os|wa|ze)|fetc|
fly(\-|_)|g1 u|g560|gene|gf\-5|g\-mo|go(\.w|od)|gr(ad|un)|haie|hcit|hd\-(m|p|t)|hei
\-|hi(pt|ta)|hp( i|ip)|hs\-c|ht(c(\-| |_|a|g|p|s|t)|tp)|hu(aw|tc)|i\-(20|go|ma)|i230|
iac( |\-|\/)|ibro|idea|ig01|ikom|im1k|inno|ipaq|iris|ja(t|v)a|jbro|jemu|jigs|kddi|
keji|kgt( |\/)|klon|kpt |kwc\-|kyo(c|k)|le(no|xi)|lg( g|\/(k|l|u)|50|54|\-[a-w])|
libw|lynx|m1\-w|m3ga|m50\/|ma(te|ui|xo)|mc(01|21|ca)|m\-cr|me(rc|ri)|mi(o8|oa|ts)|
mmef|mo(01|02|bi|de|do|t(\-| |o|v)|zz)|mt(50|p1|v )|mwbp|mywa|n10[0-2]|n20[2-3]|n30
(0|2)|n50(0|2|5)|n7(0(0|1)|10)|ne((c|m)\-|on|tf|wf|wg|wt)|nok(6|i)|nzph|o2im|op(ti|
wv)|oran|owg1|p800|pan(a|d|t)|pdxg|pg(13|\-([1-8]|c))|phil|pire|pl(ay|uc)|pn\-2|po
(ck|rt|se)|prox|psio|pt\-g|qa\-a|qc(07|12|21|32|60|\-[2-7]|i\-)|qtek|r380|r600|raks|
rim9|ro(ve|zo)|s55\/|sa(ge|ma|mm|ms|ny|va)|sc(01|h\-|oo|p\-)|sdk\/|se(c(\-|0|1)|47|
mc|nd|ri)|sgh\-|shar|sie(\-|m)|sk\-0|sl(45|id)|sm(al|ar|b3|it|t5)|so(ft|ny)|sp(01|h
\-|v\-|v )|sy(01|mb)|t2(18|50)|t6(00|10|18)|ta(gt|lk)|tcl\-|tdg\-|tel(i|m)|tim\-|t\-
mo|to(pl|sh)|ts(70|m\-|m3|m5)|tx\-9|up(\.b|g1|si)|utst|v400|v750|veri|vi(rg|te)|vk
(40|5[0-3]|\-v)|vm40|voda|vulc|vx(52|53|60|61|70|80|81|83|85|98)|w3c(\-| )|webc|whit|
wi(g |nc|nw)|wmlb|wonu|x700|yas\-|your|zeto|zte\-/i.match(request.user_agent[0..3])end
```

FIGURE 2.4 An HTTP header for mobile device detection creates a positive user experience when you are aware of the device type accessing your site.

---

[*] Chad Smith, "Detect Mobile Browsers/Open Source Mobile Phone Detection," Detect Mobile Browsers, n.d., web, n.p. (accessed May 18, 2014).

[†] "ScientiaMobile—How WURFL Works," Scientia Mobile, 2014, web (accessed October 15, 2014).

	WURFL Cloud	WURFL OnSite	WURFL InFuze	WURFL InSight
Benefits	Always updated, cloud-based device description repository Simple to install and no maintenance Affordable	Open and extensible DDR and API Locally installed for more control or integration into OEM services Highly scalable Weekly updates	High-performance C++ API network-level integration with web servers and content proxies, exposing WURFL devices' capabilities to any application across the enterprise	Integrates with existing data marts, analytics tools, weekly updates, and optimization for your web traffic
Target segment, uses	Mobile web optimization	Mobile web optimization Analytics Advertising	Mobile web optimization Analytics Advertising	Analytics
Usage	Priced by number of sites, detections, and device capabilities accessed	Single site, multiple site, and original equipment manufacturer (OEM) packages available	Single site, multiple site, and OEM packages available	Priced by capabilities accessed, application usage
Support	Forums	Support portal and forums	Support portal and forums	Support portal and forums
DDR location	Cloud	Local	Local	Local
Device capability fields	2, 5, 10, or more fields depending on package	Unlimited	Unlimited	10 or more capabilities, depending on package
No. of detections	5000, 50,000, 2 million, or 10 million, depending on package	Unlimited	Unlimited	Unlimited
Updates	Weekly	Weekly	Weekly	Weekly
API-supported languages	Java, PHP, .NET, Python, Ruby, Node.js, and Perl	Java, PHP, and .NET	C++, Varnish cache module, Apache HTTP server module, NGINX module	

FIGURE 2.5 ScientiaMobile offers four products to support businesses that aim to employ tools to gain market share based on the knowledge of user mobile device types.

2.2.2 Client-Side Detection

There is much to be learned from the user's browser and device by using feature detection. The most important thing to determine is whether or not the device has touch capabilities, and then whether the user has a large or small screen. A distinction should be made between large and small touch devices. Figure 2.6 shows both Android™ and iOS devices overlaid, including their corresponding screen resolutions. The asterisk indicates the device may come in doubled density. Even though the pixel density may be doubled, CSS will continue to specify the same sizes. CSS pixels

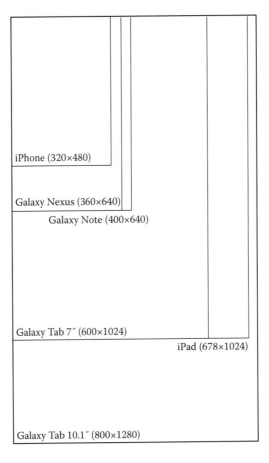

FIGURE 2.6 Android and iOS devices are overlaid to display corresponding device screen resolutions (in pixels).

via the mobile web are not the same as screen pixels. Retina® devices in the iOS platform introduced the practice of doubling pixel density (e.g., iPhone® 3GS vs. 4, iPad® 2 vs. 3). The retina Mobile Safari user-agents continue to report the same device width to avoid breaking the web. As other devices, like Android, deploy higher-resolution displays, they are employing the same device-width trick.

We must consider both landscape and portrait modes. Each time the device is reoriented, we do not want to reload the page or load additional scripts, though we may choose to render the page inversely. Figure 2.7 illustrates squares that represent the maximum dimensions of each device, as a result of overlaying the portrait and landscape outlines.

By regulating the threshold to 650 px, we classify the iPhone and Galaxy® Nexus™ as smalltouch, and the iPad and Galaxy Tab as tablet. The Galaxy Note is classified as phone, and will fall into the phone layout.

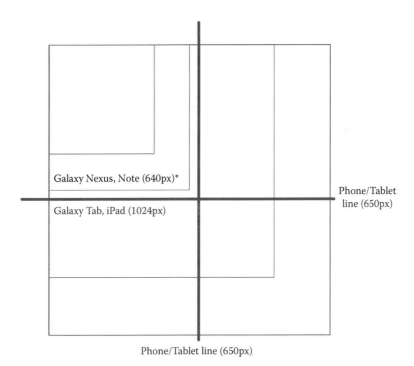

FIGURE 2.7 The maximum dimensions are shown considering both portrait and landscape modes on mobile devices.

Thus, a reasonable strategy might look like this:

```
if (hasTouch) {
  if (isSmall) {
      device = PHONE;
  } else {
     device = TABLET;
  }
} else {
   device = DESKTOP;
}
```

An alternative approach here is to use user-agent sniffing to detect device type. Ultimately you create a set of heuristics and match them against your user's `navigator.userAgent`. Simulated code looks something like this:

```
var  ua = navigator.userAgent;
for (var re in RULES) {
   if (ua.match(re)) {
      device = RULES[re];
      return;
   }
}
```

If you are performing user-agent detection on your server, you can decide which CSS, JavaScript, and document object model (DOM) to serve when a new request comes in. However, if you are doing client-side detection, the situation is more complex. There are several options, and here are two:

1. Redirect to a device type-specific URL that contains the version for this device type.

2. Dynamically load the device type-specific assets.

The first approach is straightforward, requiring a redirect such as `window.location.href = '/tablet'`. However, the location will now have this device type information appended to it, so you may want to use the History API to clean up your URL. This approach will institute a redirect, which can be slow, especially on mobile devices.

The second approach is more complicated to implement. You will need a mechanism to dynamically load CSS and JS, and (browser depending) you may not be able to do things like customize <meta viewport>. Also, since there is not a redirect, you are left with the original HTML that was served. Of course, you can manipulate it with JavaScript, but this may be slow or inelegant, depending on your application.

When deciding between client or server, there are pros and cons to both approaches.

Client:

- Scales for the future based on screen sizes/capabilities rather than user-agents.

- There is no need to constantly update the user-agent list.

Server:

- Full control of what version to serve to what devices.

- Better performance: No need for client redirects or dynamic loading.

Device.js is a good starting point for doing semantic, media query-based device detection without the need for special server-side configuration, saving the time and effort required to do user-agent string parsing. The concept is to provide search engine-friendly markup (link rel = alternate) at the top of your <head> indicating which versions of your site you intend to provide.

```
<link rel="alternate" href=" http://collins.com
"id="desktop"
     media="only screen and (touch-enabled: 0)">
```

Now you have the option to either do server-side UA detection and handle version redirection on your own, or use the device.js script to do feature-based client-side redirection.

The recommendation is to start with device.js and client-side detection. As your application evolves, and if you find client-side redirect to be a significant performance disadvantage, the device.js script can easily be removed and user-agent detection implemented on the server.

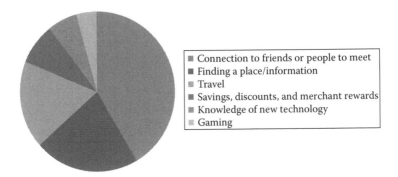

FIGURE 2.8   The most important personal use for location- and map-based apps.

2.3 LOCATION AWARENESS

Location, map-based, and geosocial apps offer a captivating experience on mobile devices. Location-based services (LBSs) obtain the geographic position of one or several individuals in an effort to create, compile, select, or filter information and serve it to the users. Users of location-based apps are drawn to such services for personal connections, such as social networking, mobile gaming, friend finder services, and navigation. The social aspect of these apps offers opportune interaction where people update their status, or *check in*, at various places they visit, e.g., local restaurant, store, airport, etc. By checking in, others know where you are, and if they are nearby, may join you. Almost half (40%) of all location-based app users are drawn to such services for personal connection, while 20% use the apps to locate places or information. Figure 2.8 exemplifies other benefits to using location- and map-based apps.

Even though over 50% of mobile users are aware of location-based apps, less than 35% are using one. Mobile users cite unfavorable reasons for not adopting location-based apps, listed in Figure 2.9. The most common implication of this technology is the concern for lack of privacy. Daily use of location-based apps is still considerably lower than that of social networking apps. Amid location-based app users, 20% check in with their favorite geosocial app at least once a day, compared to 65% who check in with their favorite social media app daily.

2.3.1 Locating Methods

Mobile devices have integrated GPS (global positioning system) tracking systems that deliver independent mobile tracking through the device,

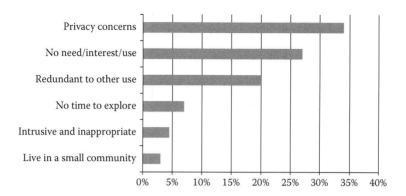

FIGURE 2.9   Reasons users do not use location-based apps, with privacy issues at the top.

alleviating the need for a separate dedicated GPS unit. The Department of Defense established GPS in 1973, utilizing a network of satellites in orbit to ascertain the location and time of a device. The data can only be sent when there are four or more satellites within line of sight of the device, referred to as triangulation. Since 2005, the Federal Communications Commission (FCC) requires mobile phones to have GPS tracking capabilities where accuracy is expected to be within roughly 100 meters for emergency purposes, referred to as E911.

Basic locating technology is derived from quantifying power levels and base station patterns, anticipating that a powered mobile device communicates wirelessly with the closest antennas. Each base station encompasses a small geographical area, referred to as a location area. Therefore, knowledge of the base station indicates the mobile device is nearby. The device itself collects its location from a variety of sources. The GPS antenna that is on the device positions itself on the globe based on satellites in the sky, and the local Wi-Fi signals allow the mobile device to triangulate and query against databases like Google and Apple®. Then the device will be told where its approximate location is; then via the cell towers, mobile providers like AT&T and Verizon allow the devices to triangulate their location and view inquiries like restaurants in their area.

Modern systems ascertain the sector in which the mobile device resides and are able to approximate the distance to the antenna, as shown in Figure 2.10. If necessary, further calculations can be done by interpolating signals between adjacent antenna towers. Mobile devices are required to relay their location at set intervals using a periodic location update

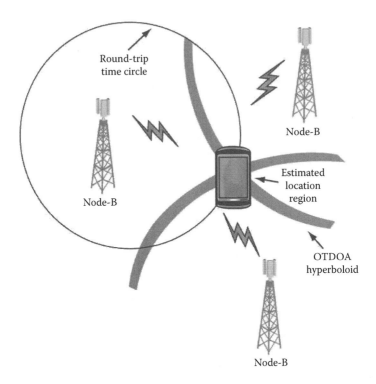

FIGURE 2.10 Cell triangulation both receives and transmits location information across three nearby cell towers. And by transmitting, the signal that the receiver receives essentially allows him or her to triangulate the location of the device.

procedure. The location update procedure permits a mobile device to notify a cellular network as it moves from one location area to the next. These updates can reduce signal fade, and the subscriber gains reliable access to the local network.

Location-based systems can be divided into the following:

- Network based
- Handset based
- SIM based
- Wi-Fi
- Hybrid

The accuracy and complexity for each location determining technology, such as Observed Time Difference of Arrival (OTDOA), Assisted Global Positioning System (A-GPS), and Cell ID, need to be integrated into the physical layer of the carrier's network. A pair of base stations defines the hyperbola, and with three base stations, the intersection or convergence can be obtained for the location of the coordinates. The coordinates can be translated to the grid in actual location coordinates the telco carrier already has stored in its location database. The content provider will output the feedback message to the requester while simultaneously making note of the total time required for a simple round-trip Short Message Service (SMS) query and service feedback message to return to the SMS requester.

2.4 SUMMARY

This chapter covers the important aspects to consider when developing and working with platforms on mobile devices. The user experience is a critical component when determining whether an app is successful or not. When considering the device and platform, the use of the device must also be assessed. Not only are these mobile devices heavily used in our personal lives, but also they are becoming more common in the workplace.

Mobile Software

Lauren Collins

CONTENTS

MOBILE OPERATING SYSTEMS, commonly referred to as mobile OS, offer a slimmed-down feature set of the widely used desktop computer. Since all functionality cannot mirror a desktop, a mobile OS must be smarter and go above and beyond the needs of power users. Thus, content is specifically developed for desktops and mobile devices, separately, with the user's experience in mind. Another component of mobile software refers to the applications running on the mobile device. Throughout this book, several flavors of software are used in examples, or referenced for both positive and negative attributes.

Mobile versions are pushed so heavily that businesses are now offering free subscriptions for online or reader versions of news and magazines. Android™ and iPhone® are the top two smartphone platforms, so developers are concentrating their efforts on serving the majority of the market on those two platforms. There is an inherent tension when resources are finite and developers must choose which platform to develop first. As a result, every app developer should be familiar with some rudimentary characteristics of both Android and iPhone, as well as other user platforms, as they are coding business applications.

3.1 MOBILE OPERATING SYSTEMS

Android, commonly referencing the robot in Figure 3.1, is the most popular operating system since the HTC Dream was released in the second quarter of 2009.* Most of Android is open source and therefore free; however, Android is also comprised of proprietary and licensed software such as Google® Play™ store, Google Site Search™, and so on. Android 2.x releases were generally for phones, but there were some tablets using 2.x. The first tablet release for Android was 3.0, and it did not run on mobile

FIGURE 3.1 Pictured is the Android robot, a symbol of the most popular mobile operating system. (Android Robot, from http://developer.android.com/distribute/tools/promote/brand.html.)

---

* "Gartner Says Worldwide Mobile Phone Sales to End Users Grew 8 Per Cent in Fourth Quarter 2009; Market Remained Flat in 2009," Gartner, n.d., http://www.gartner.com/newsroom/id/1306513 (accessed May 26, 2014).

phones. Current Android versions are nicknamed after candy or desserts: Cupcake (1.5), Donut (1.6), Éclair (2.0, 2.1), Frozen Yogurt (Froyo) (2.2), Ginger Bread (2.3), Honeycomb (3.0), Ice Cream Sandwich (4.0), Jelly Bean (4.1, 4.2, 4.3), and Kit Kat (4.4).

There are intriguing nuances behind the names for the OS versions. Not only are the operating systems named after candy or desserts, but they are released in alphabetical order. Google built lawn statues of the various Android desserts, and pictured in Figure 3.2, they reside on the lawn of Google's headquarters in Mountain View, California. Honeycomb was an OS made solely for tablets, and the latest release, Kit Kat, was originally intended to be named Key Lime Pie, but was changed as "very few people actually know the taste of key lime pie."[*]

The Apple® iOS is closed source and proprietary, and built on open-source Darwin core OS. The iOS was initially developed for the iPhone, but also encompasses other devices, such as the iPod Touch®, iPad, iPad Mini, and second generation of Apple TV®, which all run on the iOS, derived from Mac OS X.

Apple named its nonmobile operating systems after cats: Cheetah, Puma, Jaguar, Panther, Tiger, Leopard, Snow Leopard, and Lion. Maybe after Apple hits version 10, the mobile OS will be named in line with the other cats. The user interface, in this author's opinion, is the prime feature behind the success of Apple's mobile OS. The toolkit used to build the user interface, Cocoa Touch®, offers intuitive touch gestures for interface control, such as sliders, switches, and buttons for swiping, tapping, pinching, and reverse pinching.

FIGURE 3.2 Lawn statues sit in front of Google's headquarters representing some of Android's mobile operating systems.

---

[*] Leo Kelion, "Android Kit Kat Announced," BBC News, September 3, 2013, http://www.bbc.com/news/technology-23926938 (accessed June 25, 2014).

Blackberry® once held the market share for business and government usage, but has declined rapidly with the rise of cutting-edge competitors, such as Apple's iPhone and Google's Android devices. Blackberry's closed-source and proprietary devices include the Blackberry phone, PlayBook tablet, and Blackberry Enterprise Server for organizational mobile management.

With the release of Blackberry version 10, which came years after competitors introduced similar functionality, a full touch screen device was introduced. And a big surprise to anyone who has used a Blackberry, the only reason a user must press a physical button is to power the device on and off. The OS offers all the features any smartphone would: gestures, multitasking, camera, virtual keyboard, and voice control. Release 10 also introduced an Android runtime player, which allows developers to package and distribute apps.

Windows® Phone is a closed and proprietary operating system from Microsoft®. In 2010, Microsoft revealed its next-generation mobile OS, Windows Phone, which included a newly overhauled user interface inspired by Microsoft's "Metro Design Language."* Windows XP and 7 tablet versions are on a handful of devices, but Microsoft still has to do a fair amount of engineering before Windows is effective and fully functional on a tablet.

webOS is the HP® tablet operating system. webOS uses touch screen technology comparable to iPad, but application windows run in separate "cards." This allows the user to flick between different app screens, perhaps between a web page and an email, all of which are displayed on cards.

3.2 THE SECOND OPERATING SYSTEM

Most people are not aware that smartphones have two mobile operating systems. The primary, user-facing software platform (Android, iOS, and so on) is supplemented by a secondary low-level proprietary real-time operating system (Qualcomm®, MediaTek®, Infineon®, etc.), which operates the radio and other hardware. The low-level systems contain a range of security vulnerabilities, permitting malicious base stations to gain high levels of control of the mobile device. This operating system is stored in

---

* Mary Jo Foley, "Microsoft Design Language: The Newest Official Way to Refer to 'Metro'|ZDNet," ZDNet, All about Microsoft, October 29, 2012, http://www.zdnet.com/microsoft-design-language-the-newest-official-way-to-refer-to-metro-7000006526 (accessed May 27, 2014).

firmware, and runs on the baseband processor, which is always entirely proprietary. It must be capable of processing data as they come in, without buffering delays.

3.2.1 Baseband Analyzing

3.2.1.1 Case Study: Functionality of a 3G USB Broadband Aircard Modem with a Qualcomm Model Baseband (the Type Your Wireless Carrier Offers for Free with a Monthly Paid Data Plan)

When connecting this USB device to a host computer, we are able to emulate a serial line over USB and communicate with the host using standard Hayes modem AT commands, illustrated in Figure 3.3. Three entry points are initiated by plugging in the USB stick: AT commands and packet data (multiplexing), packet data (no multiplexing), and a channel for a diagnostic task.

The diagnostic channel is enabled directly on the 3G stick by initiating AT commands specific to the defined 3GPP TS specification for your baseband model. Figure 3.4 shows the diagnostic protocol, having 0x7e as both the beginning and ending markers, 1 byte for command type, and variable parameters to make up this 16-bit CRC-CCITT (cyclic redundancy check–Consultative Committee for International Telegraphy and Telephony).

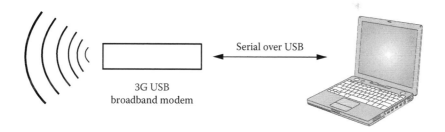

FIGURE 3.3   Once a 3G USB broadband modem is plugged into a host, three entry points are initiated.

0x7e	CMD	Variable-length data	CRC-CCITT	0x7e

FIGURE 3.4   The diagnostic protocol offers direct access to memory and supports a large range of commands.

The specification for CRC-CCITT is

- 16-bit width

- 0x1021 truncated polynomial

- 0xFFFF initial value

- Input data not reflected

- Output CRC not reflected

- No XOR (exclusive output rate) performed on the output CRC

There are a large amount of diagnostic task commands available in this instance, and all have varying purposes. Since these commands are for the baseband, the syntax does not relate to CLI (command line interface) or typical diagnostic syntax. For example, 0x02 reads a byte in memory and 0x05 writes a byte in memory. There are two options to dump the system memory: dump the primary boot loader at 0xffff0000 and dump the whole system memory from 0. A downloader mode has limited access to a small hardcoded memory range and can write data to memory or execute a command at a particular address. The downloader mode is only enabled when the system crashes, through command 58 of the diagnostic mode, or by performing the following key sequence on boot-up of some HTC phones (volUp+volDown+Power), which produces five vibrations.

The technology for this Qualcomm baseband modem dates back to 2006–2008 (which is a newer version), and runs the REX (Real-Time Executive) kernel RTOS (Real-Time Operating System).[*]

REX possesses the following features:

- Interrupt management

- Exclusive lock

- Task management and synchronization

- Preemptive multitasking

---

[*] Keonwoo Kim, Dowon Hong, Kyoil Chung, and Jae-Cheol Ryou, "Data Acquisition from Cell Phone Using Logical Approach," World Academy of Science, Engineering, and Technology, June 29, 2012, http://connection.ebscohost.com/c/articles/28543399/data-acquisition-from-cell-phone-using-logical-approach (accessed May 27, 2014).

3.2.2 Boot Process

Qualcomm's operating system is called Advanced Mobile Subscriber Software (AMSS), also referred to as Dual-Mode Subscriber Station (DMSS). The OS allocates resources for up to 69 concurrent tasks, including hardware management for USB, USIM, GPS, etc., and contains protocol stacks at each layer (GSM, L1, L2, RR, MM, etc.). The primary processor boots first, ARMx, executing the primary boot loader (PBL) from on-board ROM at 0xFFFF0000. The mobile station modem (MSM) platform design couples CDMA (code-division multiple access) and UMTS (Universal Mobile Telecommunications System) modem chipsets. After hardware initialization, the PBL reads the device boot loader (DBL) from the first partition of flash memory. DBL is a function of Qualcomm's SecureBoot, which uses cryptography to guarantee there has not been interference with the boot loader images. DBL then configures a dedicated cryptographic co-processor, Cryptographic Look-aside Processor (CLP), and other hardware to load and execute the secondary boot loader (SBL) from flash memory on an external bus interface (2) from a dedicated partition (3).

The SBL is loaded into memory, which is an internal memory location, on the MSM internal memory, the package-on-package RAM. This is the ARMx Monitor (AMON), where x signifies the primary processor model number, and it provides an Extensible Firmware Interface (EFI) environment for controlling the boot process. After more hardware configuration takes place, including USB and universal asynchronous receiving/transmitting (for potential remote console connections to the monitor), it loads the Applications Processor Secondary Boot Loader (hboot) on the ARMxx applications processor from a dedicated partition (14) into both virtual and physical memory.

REX and AMSS are loaded and executed from another dedicated partition (4). REX, running on Security Domain 0, is responsible for loading firmware into the ancillary micro-controller, digital signal processor. and voice processor and initializing them. The image includes the REX-embedded micro-kernel, such as L4A Pistachio, and Iguana operating system combination, with extensive Qualcomm modifications and extensions. Once ARMxx boots, REX unloads/disconnects its eMMC (multimedia card) driver and relies on remote procedure calls (RPCs) via shared memory to the ARMxx application processor to read and write the eMMC. Subsequently, the ARMx REX executes the Advanced Mobile Subscriber Software (AMSS), running on Security Domain 1.

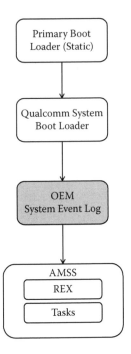

FIGURE 3.5 Boot sequence of how the Qualcomm mobile station modem system-on-chip boot-straps the processors into a mobile OS.

Figure 3.5 illustrates the flow for the system boot sequence.
Core REX tasks are

- DIAG: Provides the diagnostic interface.

- CM: Call manager task.

- DS: Data services task (unified data gathering task for all protocol layers).

- DOG: Watchdog (constantly checks whether tasks are alive).

- DPC: Routes adjacent point code (APC) across tasks.

- MAIN: Launches all system tasks, handles timer events.

- PS: Packet-switched services (network stacks at upper layers, TCP/IP, PPP, etc.).

- SLEEP: Idle task.

3.2.3 HBOOT

If you are looking to reverse engineer the boot image on your mobile device, there are open-source code mappings available on the Android source code repository. HBOOT is a boot loader that lives inside the first partition and is generally used by HTC devices. This secondary boot loader is responsible for configuring, loading, and passing execution to the applications operating system by checking the hardware, initializing the hardware, and starting either the Android or Recovery OS.

When using HBOOT commands over the serial interface, use the `screen` utility to establish a connection to the port and communicate with the device.

Notice: Be sure to read the risks associated when issuing commands in HBOOT. The author is not suggesting or recommending these commands be performed.

Possibly just as important as the notice above, there is a special key sequence that can be performed to exit this mode:

```
screen/dev/ttyUSB0
```

Even though the screen is blank, enter the above command and press enter:

```
hboot>
```

Here is a brief list of some commands:

- `battcheck <param1>`
  `//battery check`

- `bdaddress <param1>:<param2>:<param3>:<param4>:<param5>:<param6>`
  `//set Bluetooth address (Bluetooth MAC)`

  `emapiCountryID`
  `//display country signature and ID`

3.3 APPLICATION PLATFORMS

How does a developer perceive different platforms? How are platform choices affected by the type of apps developed? What is a developer's definition of a successful app? Ideally, all supported platforms should be allocated the same level of resources required in order to deliver the best possible product. In business, however, this is rarely the case, particularly for large organizations.

Every application relies on other software: operating systems, proprietary software, database management software, and even software running in the cloud (software as a service (SaaS)). No matter where these multiple software systems are running, all of them intersect to reference an application platform, running on your mobile device. Applications are made up of data sets, and those data sets represent extreme value to people, businesses, and technology. Every application that is installed, referenced, or used depends on the application platform. An effective application platform is responsible for delivering appropriate services for every device, including mobile phones, tablets, desktops, local servers, and cloud servers. Each type of application requires different resources, interfaces, and accessibilities from an application platform. A mobile device employs a single-user application that uses completely different services for storage and execution than an application would that is running in the cloud supporting a few hundred users. Considering both the conceptual use and device of an application presents an expansive view of the application platform, the services it will provide anywhere the application will run.

3.3.1 Content Delivery

Wireless Application Protocol (WAP) is a specification for a set of communication protocols that standardizes the way mobile devices can access the Internet. While Internet access was possible before, multiple manufacturers supported different technologies. Service providers had limited options for their customers, but were able to support:

- Email

- Real-time stock market quotes

- News and sports headlines

- Music downloads

As mobile technology continues to play an integral part in our lives, consumers are now realizing the potential of their mobile devices extends far beyond making calls and sending text messages. Consumers look to do more on the Internet, have the desire to work remotely, and are eager to experience a variety of content on their mobile devices, immediately. To instantaneously and effortlessly deliver content, On-Device Portal (ODP) solutions were developed and are now deployed everywhere.

Users can now easily and intuitively view, record, or exchange content on a mobile device. ODP is a step up from WAP, because device capabilities are leveraged to deliver a more compelling user experience, increase service awareness, and streamline content delivery.

In Figure 3.6, WAP, ODP, and web portals are the interfaces between users and content. Web portals are still the only full-featured version

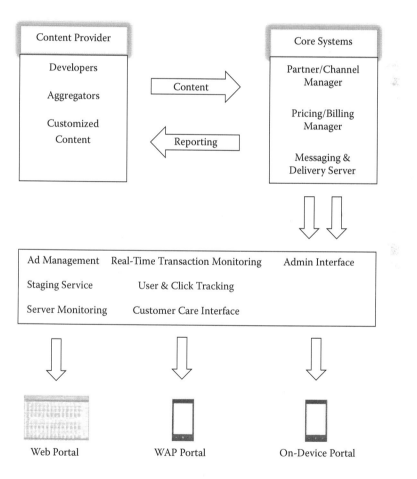

FIGURE 3.6 Mobile devices support three different portal experiences. On-Device Portal leverages mobile device capabilities to deliver a more compelling user experience and streamline content delivery.

of any complete shopping experience or software application; however, mobile technology must support the lifestyle of people who are constantly moving.

On-Device portals include a categorized and personalized menu structure, one-touch access to mobile services, and quick links to WAP portal sections. ODP not only allows for the provision of services via mobile, but also promotes the analysis and feedback of adoption and usage that supports the development of services that users actually want to use and share with other users.

3.3.2 Benefits of Content Delivery

On-Device Portal drives revenue and service usability by delivering content directly to a mobile device in an easily discoverable, accessible, and personalized format. This has become so vital to businesses and consumers alike that today entire business units focus on content delivery. Here are some additional business values to consider regarding the format to project content delivery:

- Delivery of focused mobile services
  - Add click-to-call and instant messaging capabilities
  - Review website processes to make them mobile-friendly
- Customer loyalty by improved branding
- Personalization
- Track and report user behavior
  - Increase download sales and data revenue by improved usability

Content delivery is important to users where flow and efficiency are concerned. Stores and businesses entice us to download their app when searching and shopping, and the user experience is more effective and positive than visiting their website.

3.4 DEVELOPMENT ENVIRONMENT PLATFORMS

As market needs for cross-platform tools continue to rise, it is clear why we are seeing a mobile fragmentation that is much larger and more significant than the recent similar crusades waged over the desktop. This

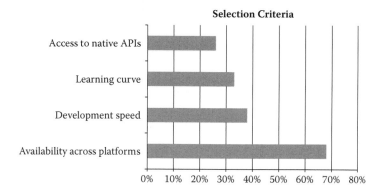

FIGURE 3.7 Top reasons developers use cross-platform tools to develop mobile applications.

fragmentation is telling us that cross-platform tools (CPTs) for mobile development are not a one-size-fits-all solution.* Figure 3.7 illustrates the selection criteria used by developers when selecting their CPT.

3.4.1 PhoneGap

The Adobe® PhoneGap™ is a mobile development framework perfect for a beginner that permits the developer to build an app that is capable of running on multiple device types. PhoneGap supports the following OSs: Android, iOS, Blackberry, webOS, Windows Phone, Symbian, and Tizen. The tools used within PhoneGap bundle the HTML, CSS, and JavaScript files into platform-specific deployment packages. JavaScript is utilized for the logic, and HTML5 and CSS3 are utilized for rendering. PhoneGap is not a tool to assist you in building the user interface, but affords the developer tools relative to things like the file system and geolocation.

As with any tool, there are several well-known issues when using PhoneGap. Many developers have reported a click delay, or issue with screen interaction, and there is a script to run that may fix this issue: FastClick.js. Of course, the nature of your app will present many challenges, as will the platform chosen to develop with. Additionally, the use of plug-ins to perform background tasks should always be addressed with your application's background service layer.

---

* Jim Cowart, "Pros and Cons of the Top 5 Cross-Platform Tools," Developer Economics, November 12, 2013, http://www.developereconomics.com/pros-cons-top-5-cross-platform-tools.

3.4.2 Appcelerator

Appcelerator® also offers cross-platform compatibility utilizing web technology. Native apps are rendered on the mobile device via browsers and derived from HTML5, CSS3, JavaScript, Ruby, PHY, and Python. There are hundreds of APIs offered to users, and beginners are welcome to take the Beginning Appcelerator Titanium course.

3.4.3 RhoMobile

RhoMobile™ provides an integrated framework used for creating apps across various platforms, and is constructed on Ruby. RhoMobile works with iOS, Android, and Windows, and is built to sustain future operating systems. The RhoMobile Suite either complements the enterprise-wide business solution or will serve a small operation with one of its applications. Another plus to this suite is the hosting and cloud emphasis, which is a huge part of business today.

3.4.4 Xamarin

Everyone has either heard of this platform or used it. Code is usually supporting Android or iOS, written in C#, and executed on the .NET framework. Businesses that have mobile teams employ Xamarin®, as different teams are able to successfully distribute the project accordingly. Xamarin integrates with Visual Studio, which also supports building, deploying, debugging, and the simulation of a device. Xamarin is also able to integrate cloud services, libraries, UI controls, and back-end systems directly into mobile apps.

3.4.5 Bootstrap

The author's favorite, Bootstrap®, is a perfect open-source, front-end framework for any skill set, from beginner to a rockstar developer. Bootstrap, formerly known as Twitter Blueprint, was created by a developer/designer who worked at Twitter. During a hacking competition, developers of all skill levels were able to hit the ground running, and from then until now Bootstrap continues to serve as a design and style guide for Twitter, among other companies, since its public release in August 2011. Let's say you are looking for a wrapper to get off the ground quickly; WrapBootstrap has an endless supply of layouts, backgrounds, forms, and templates.

3.5 SUMMARY

As a user, selecting a mobile device OS is usually based on the look and feel, functionality, and a positive user experience. Businesses may select mobile OSs on whether or not particular applications and programs are supported on devices. And as a developer, it is important to select a tool that coincides with the overall project plan for the particular operating system(s).

CHAPTER 4

Enterprise Mobile Strategy

Lauren Collins and Charlie Connor

CONTENTS

Businesses and entrepreneurs continue to investigate ways to extend the capabilities of their brand by using mobile technology to reach their consumers more than ever. It is clear that people are on their mobile devices more than ever, and businesses want to be where their customers are. Businesses also want to be where they can have real-time impact on their customers' buying behavior. In a recent presentation from Mary Meeker about 2014 Internet trends, the data are clear that mobile is here to stay and the mobile user will become more sophisticated. For example, shown in Figure 4.1, Meeker points out that only 30% of the 5.2 billion mobile users are on smartphone.[*]

She also points out that tablets are growing faster than PCs ever did, with a 52% year-over-year growth. You do not have to look hard to see these numbers or see the efforts of competition in this space. You just can look around the next time you are riding the train or on a busy sidewalk: everyone is staring at their phone. So if the customers are there, why do so

---

[*] Mary Meeker, "2014 Internet Trends," Kleiner Perkins Caufield Byers, May 28, 2014, http://www.kpcb.com/internet-trends (accessed July 6, 2014).

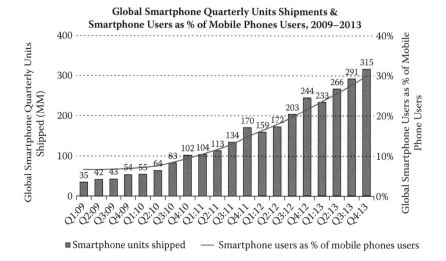

FIGURE 4.1 Smartphones only account for 30% of the user base for mobile phone use. (From Meeker, M., "2014 Internet Trends," Kleiner Perkins Caufield Byers, May 28, 2014, http://www.kpcb.com/internet-trends (accessed July 6, 2014).)

many companies fail with such a huge growing market? The answer can be broken down to two parts:

1. Do you have a mobile application or a mobile strategy?
2. An app that does it all does nothing at all.

What do we mean by mobile application vs. mobile strategy? Unless you are in a niche market or have only one engineer who can code a single language, you are going to be hard-pressed to choose to only write for a single mobile operating system. Multiple companies, from Facebook®, Microsoft®, and others, have failed on the one-size-fits-all HTML5 design. While there are short-term fillers, such as PhoneGap and Bootstrap 2.0, there is still no better choice than native applications. In a podcast, "Crafting Elegant Solutions,"* Gentry Underwood spoke to completely rebuilding Dropbox differently for each type of device its product could be used on.

"An app that does it all does nothing at all." As Mary Meeker sums up perfectly in her trends report, we are seeing the next evolution. First, it was multipurpose websites, then multipurpose applications. In particular, mobile apps serve a single purpose. Think about how you use your phone,

---

* "Gentry Underwood—Crafting Elegant Solutions," SoundCloud, https://soundcloud.com/dormroomtycoon/gentry-underwood (accessed June 30, 2014).

use your camera, check email, check the map, and text. Those are all apps, but they come to your phone. This convenience and usability is why you see the fight for the phones to be the top original equipment manufacturer (OEM) and run on the best OS. This is also why you are seeing the industry cheer over companies like Facebook. Facebook's development cycle successfully integrates multiple mobile functions within a single app. Consequently, the fast followers are not far behind.

4.1 EVOLUTION OF MESSAGING

Social strategy is now an important element to integrate within an app, whether it is chat, text capabilities, or an all-encompassing product. The top activities on a smartphone are text messaging and social networking, so consider the business roadmap for Facebook. To stay ahead of the technological trend, Facebook acquired a product, Instagram™, which has the ability to instantaneously capture and share moments. Next, Facebook invested in WhatsApp, a messaging app that will also have voice capabilities. The telecommunications industry will need to reconsider discontinuing limitless data with the implementation of WhatsApp.

Facebook has catered to mobile device users magnificently. There are three ways to access Facebook on a mobile device: mobile text message, mobile uploads, and mobile web browsing. Figure 4.2 shows the shift in messaging trends from utilizing social media to broadcast a message to a

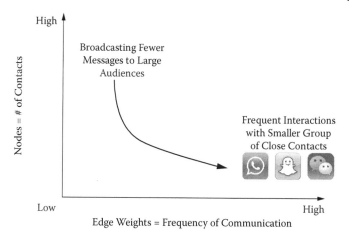

FIGURE 4.2   Trends in the size of social media contact base are moving toward more frequent communication with a smaller set of contacts utilizing messaging apps. (From Meeker, M., "2014 Internet Trends," Kleiner Perkins Caufield Byers, May 28, 2014, http://www.kpcb.com/internet-trends (accessed July 6, 2014).)

large audience (friends) to employing frequent interactions with a smaller group of contacts. More personable and meaningful messages can be communicated in the form of photographs or videos to a smaller audience, expecting the audience to both immediately view the content and comment.

On a higher level, when sending a text message from your mobile device to Facebook, the message transmits to a mobile switching center (MSC), which sends the signal to a signal transfer point (STP). Then, the message goes to a short message service center (SMSC), which sends the text to Facebook. When Facebook sends a text to your mobile device, the process is reversed. Using text messages, you can look up basic member profile information, send messages, add friends, and interact with some Facebook applications.

Mobile uploads are similar to text messaging, but use Multimedia Messaging Service (MMS). MMS allows the user to send a text, audio files, video, and images. The transfer method is similar to Short Message Service (SMS), but requires a handheld device compatible with MMS. Since there are a handful of mobile devices that are not compatible with MMS, service providers build in a feature alerting the user when he or she has received a MMS message. The message instructs the user to visit a web link to view the message. A MMS-compatible device allows you to upload photos to your profile. They will appear in a special uploaded photos section. Your mobile device will also allow uploaded notes or videos to your profile. In either case, the MMS message is created and then sent to the proper address. Figure 4.3 shows the interface of a multipurpose web app on the left, and how the evolution of app functionality supports multipurpose

FIGURE 4.3 Pictured from left to right, web application interfaces present too much information for the mobile user and do not fill the screen for the user to easily view or navigate. Mobile devices should provide the user with the look and feel of an app, yet not unbundle functionality.

mobile apps on the right. The composition of a mobile device dictates the need for an interface to be intuitive, yet less complicated for positive user experience and improved performance.

4.1.1 Instagram

The saying "A picture is worth a thousand words" is precisely the direction Facebook was headed in with the purchase of Instagram, a photo-sharing network. The perception is that Facebook invested in another social network piece, but considering the marketing and advertising success for companies, this is an obvious target for success in monetization. Prior to Facebook's 2013 purchase of Instagram, General Electric was avid in marketing its new technology and engineering. Pictured in Figure 4.4, the advertisements on Instagram show and feel more organic than the ads from other social networks. Instagram has not only made a name for itself, but has uncanny momentum around it, including a lot of growth potential in the future. Beyoncé posted an advertisement for Pepsi with over 300,000 likes, and Figure 4.5 shows Kim Kardashian's post with over 2 million likes.

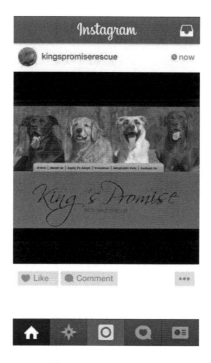

FIGURE 4.4 Instagram is a social networking app supporting monetization for products, services, or events.

FIGURE 4.5 Kim Kardashian posted a photo with the all-time most likes on Instagram, over 200 million.

Instagram allows individuals and brands to market and promote themselves on a native platform. Figure 4.6 shows the initial process of downloading the app from iTunes, and how either an email or Facebook account can be tied to Instagram. Either pictures or videos can be uploaded to Instagram and propagated to your followers. The user can include captions for uploads or simply post the object and await his or her followers' likes and comments.

The media is still wondering when Instagram will include a component for shopping or a way to make money outside of advertising. Users and other companies have found alternate means to benefit their needs. For example, if your follower count is low, there are multiple companies where you can purchase Instagram likes and followers. Currently Instagram is only supported on Android™ and iPhone®, but Instagram.com is also an option.

4.1.2 WhatsApp

In the beginning of 2014, Facebook purchased yet another social media network, WhatsApp®. While WhatsApp allows users to share photos and videos similar to Instagram, users can also send real-time messages directly or as a group. The capability even exists to send a voice message to one of your contacts. Supported platforms include Android, Blackberry,

FIGURE 4.6 Pictured left to right, Instagram can be downloaded from the App Store; an Instagram account can be tied to either your e-mail or Facebook account.

Nokia, iPhone, and Windows® Phone. The founders of WhatsApp were very deliberate to keep ads from being a part of their platform. The app is free for the first year, and then a yearly subscription fee applies. Considering the amount of data messaging and photos may accumulate, the subscription price is nominal, especially after a year of free use.

4.2 CONTENT AWARENESS AND SERVICING

Mobile devices are built for convenience, speed, and to provide an enhanced user experience. Thus, applications that are resource-intensive are no longer performing operations on the mobile device. Businesses achieve content awareness with technologies using inspection techniques and mechanisms, such as precise data matching, statistical analysis of data, keyword and file tagging lookups, and watermark recognition. For the purpose of this chapter, operations such as streaming music and rendering photos are performed on cloud servers in an effort to determine the information contained on a device, which also ends up conserving the battery and CPU. Network factors must also be taken into consideration when determining an approach to content awareness. For example, if a photo is rendered in the cloud and is then streamed to a mobile device over a wireless network with intermittent connectivity and limited bandwidth,

FIGURE 4.7 Pictured from left to right, content-aware tools are supported on mobile applications to render cropping, apply filters, and create other effects.

the user experience suffers considerably. Ultimately, once determining and prioritizing factors, the compromise between user experience and content awareness can be measured.

Adobe® released Photoshop® Elements 12 in 2013, which supports mobile photo applications utilizing content-aware tools. Instagram-inspired effects are features of Adobe's editor, including textures, effects, frames, new guided edits, and content-aware move tools. Figure 4.7 shows the various options offered when rendering a photo of a cat in Instagram.

Animal lovers are also able to use the Elements 12 image correction feature that remembers how a red-eye tool was used on a dog or cat, a feature that Photoshop does not have. Another attractive feature Adobe realized is the challenge presented in removing images off your mobile device. Adobe Revel is a free online storage, syncing, and sharing utility that is similar to the Apple® iCloud and Photo Stream. Photos can be conveniently uploaded into Revel or with the desktop version Elements 12, shared automatically, synced between the applications, and shared via email.

4.3 SUMMARY

Facebook has done a tremendous job assessing the market trends and staying well ahead of the competition. When Facebook was introduced, using the product on laptops was cutting edge. As technology has shifted to the use of mobile devices, Facebook has made its products more user-friendly for mobile devices and developed its own platform apps for photos, messaging, and a magnitude of new features. We are in the era where apps are deployed in layers. The evolution of apps is built with a purpose and is informed by hardware sensors, location, history of use, and predictive computation.[*] Anyone with a mobile device is capable of providing searchable and sharable data based on not only what was accessed but also location. The positive outlook is that the volume of data offers the potential to yield patterns to help solve basic or previously impossible problems, but also creates new challenges related to privacy and rights. Developers are finding consumer patterns using data mining and analytics tools, which allow for the prompt emergence of data and pattern-driven problem solving.

---

[*] Matthew Panzarino, TechCrunch, May 15, 2014, http://techcrunch.com/2014/05/15/foursquares-swarm-and-the-rise-of-the-invisible-app (accessed July 6, 2014).

II

Designing Mobile Apps

App Strategy

Jonian Musa

CONTENTS

S TEVE JOBS ONCE SAID, "Design is a funny word. Some people think design means how it looks. But of course, if you dig deeper, it's really how it works." Contrary to what most people think, the designers of an app are not solely responsible for the look and feel of the product, but also for the interface between the users and the product.

As discussed in Chapter 4, when developers are armed with knowledge, they are in a better position to choose the right development tools for a mobile application's needs. When coworkers have experience in something, it is imperative to tap in to their skill set. Not only will this save a ton of time, but they will be able to provide some insight into how certain tools are utilized and will be able to highlight successful practices. Aside from your friendly coworkers, there is a wealth of free online information available regarding code sets and tools to use for specific projects.

Developing cross-platform apps can be very challenging, as each platform is unique and exhibits unique features, proficiencies, and mannerisms. Developers would traditionally design apps for each platform independently, but this adds to the development cost, something that smaller businesses cannot afford monetarily. And larger businesses cannot afford to spend the time. Luckily, there are many tools available to developers that make cross-platform development and porting effortless. These apps not only make the job easy, but also have the added benefit of reducing the turnaround time.

The following tools represent different flavors supporting all mobile platforms:

RhoMobile: RhoMobile comes with Rodes, which is an open-source framework based on Ruby. You will be able to create impressive apps that are compatible over a range of devices: Android™, Windows® Mobile, Symbian®, iPhone®, and RIM®. Native apps for all the smartphone platforms can easily be built over this platform, which serves as a great advantage.

Unity 3D: As the name suggests, this tool is used to develop 3D apps or 3D games for mobile devices. Unity 3D has a narrow learning curve, which makes its use very prevalent among mobile developers. It can create apps for iOS, Android, and the web. Unity 3D is a visual tool that helps to reduce the amount of coding that goes into creating an engaging app. Simply design the apps by dragging objects from one to another and immediately view the graphical representations of your objects in 3D. This tool works with scripting languages such as JavaScript, C#, or Boo.

Gideros: If you are looking for a tool that will develop apps for Android and iOS quickly, Gideros is the tool for you. Gideros comes with an integrated development environment (IDE) and a built-in player that facilitates testing of the apps alongside the development process. Here you will be able to test your app across different resolutions and screen sizes, something that is necessary while developing apps for smartphones and tablets. You will also be able to test your app wirelessly for iOS and Android devices, which has proven to be a game changer.

Corona SDK: Corona SDK is one of the most popular tools for building cross-platform mobile apps and games. It uses a scripting language, Lua, which is simple for any developer to grasp. Corona SDK comes with a robust text editor, which allows you to develop apps for Android, iOS, Kindle, and Nook. Windows platforms are coming soon as well. It also makes game development easy with the Box2D physics library. This allows you to develop a wide range of games. The only downside of this tool is the fact that you are required to upload the project in its server for compilation. This is unlike any other tool, because it allows you to do everything in your environment.

5.1 DEFINE THE GOAL

The creation of anything comes after discovery, an idea spawns collaboration. Whether a team is given a technical briefing, a design summary, or simply an objective that has yet to be discussed in person, the relevant project tasks are assigned. Whether the app is developed for business or personal use, there is a flow of tasks that need to occur in sequence. The flow of tasks is as follows:

- Understand the details: It is critical for a team to understand the business or personal goal you are attempting to achieve for a specific project. Each project will have goals, and some may even have business unit-specific goals. Whether the goals are to push out a mobile version of a product or integrate mobile capabilities into your brand, it is critical to understand why you are working on the app—the design will be tied to the goal.

Designing an app is a difficult undertaking. However, it is not the most difficult part in the project plan of creating an app. The most difficult part on the path to creating an application is the conception of the application itself. The idea to create a successful app can come from a single moment when a person thinks of something special, months of hard work brainstorming about what app to create, or when a need is recognized for the work you are currently doing and realizing that creating an app will take your business to the next level.

5.1.1 Competitor Analysis

There is no doubt you have already found the most popular, competing app to yours. Or maybe your idea evolved from an app that is out there but just does not have all the bells and whistles you will integrate into your app. Throughout the process of designing an app, continue to compare products and user's opinions in various markets. Spend some initial time parsing through the app store where your applications will be published.

The commentary and ratings of real users will be an invaluable guide to move you in the right direction.

- Target audience: If the team puts enough thought in analysis of the product's audience, they will save a lot of errors and may even impress the audience. For instance, if your app is targeted toward business professionals, a flashy app with cartoon design is probably a bad idea. Likewise, a clean-cut design will not draw children in.

Let's begin with an example of creating a new application. As mentioned above, sometimes these come as a flash of genius, but most of the time these ideas come from a brainstorming session. A few years ago, a group of friends and I determined that by sitting down for a few hours we could come up with a noteworthy project. We were young and inexperienced; we did start in the right place. The meeting did not start with what we thought was a cool idea. Instead, it began with us listening to current businesses with a strong presence in the market that we thought could be improved upon or at the very least imitated. Finding an already established business and following its business model or creating something that will improve upon what said business has already had success with will give you and your team a detailed description of your target audience.

The target audience is the target market for your application. These are the people that you are aiming to sell your application to. For example, if you are selling video games, your target market is most likely 8- to 24-year-old men. When defining a target market, it is important to define the demographic, the need the demographic has for your product, and the delivery method by which you will sell this product to the target group.

5.1.2 User Base

The qualities that define your target market are demographics. Demographics include gender, age, geographic location, language, marital status, and class. Figure 5.1 portrays statistics gathered by app marketers who correlate app downloads with user satisfaction.

- Compose use cases: Consider the most popular use for your app. At this stage, think about the persona and how this persona will be using the app. This investigation will uncover a myriad of essential details, which will benefit both the design and the functionality of the app.

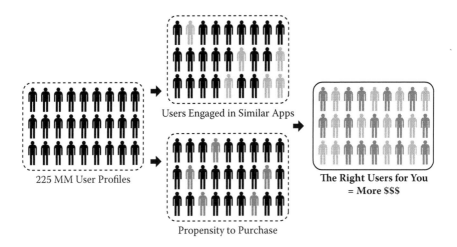

FIGURE 5.1 When designing and developing an app, be sure to target the proper audience.

Continuing with the above example, the target audience we chose to go after at the time was the market of social media—this being Myspace (yes, this was quite a while ago) and Facebook. This market is very large and growing at a rapid pace. Think of applications such as Myspace, Facebook, LinkedIn, Instagram, and Snapchat. Each of those applications has expanded or gone after a certain market type in order to be successful.

While it can be argued that the demographic for the target market of a social media market can be anyone, we decided to come up with an idea to go after the younger crowd, the trendsetters, as it were. Every day we would see more and more people on Facebook share pictures of everything. It had soon become a trend to share pictures of everything from a person's daily life, including what types of eggs a person had for breakfast.

The action of these social media sites overflowing with pictures made us think that the next logical evolution for this trend was videos. The idea was to use this young trendsetter audience to begin posting videos of their many activities that they would like to share with friends, instead of pictures. By using an existing target audience and evolving an already good idea, the idea of sharing pictures, the team was very confident that the idea would be successful.

When deciding to enter a market with a target audience as large as the social media target market, it is best to first aim for a smaller group. As shown in Figure 5.2, the annual household income share is repre-

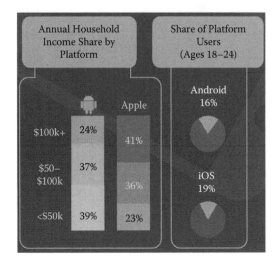

FIGURE 5.2 The left-hand side represents the annual household income divided between the platforms Android and Apple. The right-hand side displays a smaller user base of 18- to 24-year-olds, which proves a hot market to target for app rollout.

sented between Android and iPhone users.[*] Consequently, to the right in Figure 5.2, a smaller market share is represented by the young trendsetters: 18- to 24-year-old males and females will quickly spread the word to their friends if they find an application to be noteworthy enough. Then there are middle-aged adults that are looking for an easier way to maybe stay in touch with their colleagues or find previous associates from college. Lastly, there are gamers who are looking to connect with their friends via the different games that they could play on social media sites.

Other groups in this market include environmentalists, concerned parents, workout enthusiasts, health advocates, etc.

The next logical step after defining a target market and designing an application is to advertise. Once you are confident that you have a good target market and that your application is production ready, you must put together a strong marketing strategy.

5.2 MEDIA STRATEGY

With the coming of smartphones, mobile applications also came. We have come to an age where people are constantly with their head down, staring

---

[*] Benjamin Travis, "Android vs. IOS: User Differences Every Developer Should Know," ComScore, March 6, 2013, http://www.comscore.com/esl/Insights/Blog/Android-vs-iOS-User-Differences-Every-Developer-Should-Know (accessed June 1, 2014).

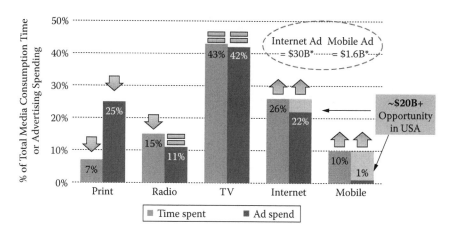

FIGURE 5.3   Mobile marketing is on the rise.

at their phone, so much in fact that chiropractor appointments have gone up a whopping 45%.

People spend 35% of their day working, 35% of their day sleeping, 8% of their day commuting, and about 22% of their day on leisure activities and hobbies. Assume that working and sleeping are untouchable hours. The untouchable hours are the hours that you cannot reach the client with a marketing message. Even though those untouchable hours come out to about 16 hours, there are still 8 hours in a day that your target audience is ready to receive your marketing message. As a business owner, these are 8 invaluable hours that you cannot spend sitting idle.

Figure 5.3 illustrates the relationship between *time spent* and *ad spend*, proving mobile marketing is the future.* Suppose, for example, you own a pizza restaurant, you make delicious pizza, and your clients love the restaurant. Your business does great on the weekends, but not so great on the weekdays, even though you have a great client base.

How is this possible? The weekdays are some of the days that food businesses are meant to thrive. People are just finishing work; by the time they have left the office and gotten to the train, their fatigue has overcome their resolve to work out and they are even too tired to cook.

The commute home from work is prime time for delivery and pickup food businesses. Without an online presence here, as a business owner you are losing valuable revenue. On the commute home from work, take a look

---

* Nicholas Carlson, "Mary Meeker's Latest Stunning Presentation about the State of the Web," Business Insider, May 30, 2012, http://www.businessinsider.com/mary-meekers-latest-incredibly-insightful-presentation-about-the-state-of-the-web-2012-5?op=1 (accessed June 2, 2014).

around and see if anyone is not staring at a phone or tablet. This 8 hours of leisure and commute that people have during the day is mostly spent on technology.

What about calling? If people truly love your pizza, then they would have the phone number and call. It is difficult to sell change and software to old-fashioned people (the telephone was invented in 1833, so yes it is old-fashioned), but how loud is the train or the bus? Would you rather use the phone to make a call or order something quietly via an application? Or how about when one is at work? Is it better to make the order quietly or call during work hours for personal business?

It is clear from the actions of people today that using an application or texting is more preferable than calling. To survive in this age, every business should have an online presence. With the booming tech industry, even old mom-and-pop shops have taken to the online community.

Creating applications nowadays is an easy service that can be provided by many companies. It would be unthinkable as a business owner not to take advantage of the time that your clients have during their idle hours. With the amount of time people spend on the Internet, you are losing money as a business owner by not providing access to your target audience via the Internet.

5.2.1 User Dynamics by Platform

The type of phone a user has represents the characteristic that makes it attractive from a development standpoint. Android users have broader content category reach, despite iOS users' higher propensity. Obvious differences between platforms are not only the way an interface looks, but also how it engages with mobile content. iPhone users are oftentimes referred to as *power users* on their phones and are more likely to engage in all major content categories than Android users. While iPhone users engage more with content on average, the Android platform has a greater number of media users in each category. This poses an interesting dilemma for developers who must consider whether audience size or engagement is the more important determinant of success for their apps. When choosing between having a higher number of users and having the type of users who engage more frequently, the decision will depend on how the developer intends to monetize the apps.

Device satisfaction and brand loyalty are differentiating factors that marketers, strategists, and developers also must consider since they, in part, determine long-term market share. iPhone users tend to think very highly of their devices, and as a result, they are likely to remain iPhone users over time. Since the majority of Android phones' hardware is not Google-branded, Apple's tightly linked hardware and software may offer a certain branding advantage that provokes higher loyalty. Of course, Apple also has a history of excellent marketing centered on the brand's iconic status.

5.3 PLATFORM MECHANICS

Once the idea for the app has been decided upon and how the app would function is decided, it is time to choose which platform to first release it to. Each different mobile platform has different coding strategies. Down the road, depending on the resources available and the success of the app, there will be time to develop on multiple platforms, but first, in order to gain a hold of the market and to retrieve some finances for further development, you will need to choose one platform to first release to. The most popular of the platforms are iOS, Android, Windows® Phone, webOS, and Blackberry®. There are of course many other software platforms for apps, such as Firefox OS, Sailfish OS, Tizen, Ubuntu Touch OS, etc. This book will discuss the five most popular, mentioned above, as they are currently the platforms that have a solid hold on the market.

Now that the list of mobile software platforms has been whittled down to five, it is time to research further and choose one to begin developing on. The most important thing that will drive the decision for the platform is to know your market. Who are you developing this app for? Once that question has been answered, then it is time to move forward with choosing a platform.

There have been studies that prove that certain genders and ages use different devices. Again, all this comes back to who is your target audience. They are the people that will ultimately be using this application.

Take, for example, the application in question is one that can map geographic locations of your favorite fishing spots and the type of fish that one can catch there. The target market research that was done prior to this shows that your target market is middle-aged men. Shockingly, at the time the most preferred mobile device used by middle-aged men that are active in the smartphone market is the Blackberry.

As you sit down and begin coding the new application, a friend walks in and you notice that he has not picked up his head from his phone. Your friend begins to reveal how his company has recently made the switch from Blackberry to mobile devices that use the Android platform.

This friend of yours works at a large company that employs middle-aged men, and this company was the initial release of the application using your close friend as an in.

While the target audience largely influences the decision for which platform to use when developing an application, it is not the sole decider. There are other considerations that must be explored to make the application as future-proof as possible.

To choose a mobile networking platform a few items must be considered:

1. How much of the market does each of the different platforms control?

2. What are the futures of these markets? Research past and current markets. Which of these platforms have the highest chance of increasing their market share and ultimately increasing the return on your investment?

3. Which platform is easiest to develop on? Which platform can you secure a good resource to develop the app?

Market research will show that the market is largely dominated by Android and iOS, with Windows, Blackberry, and others sharing the rest of the market, shown in Figure 5.4.

There is a wide margin of difference when it comes to control of the market between iOS and Android and the other platforms. As mentioned

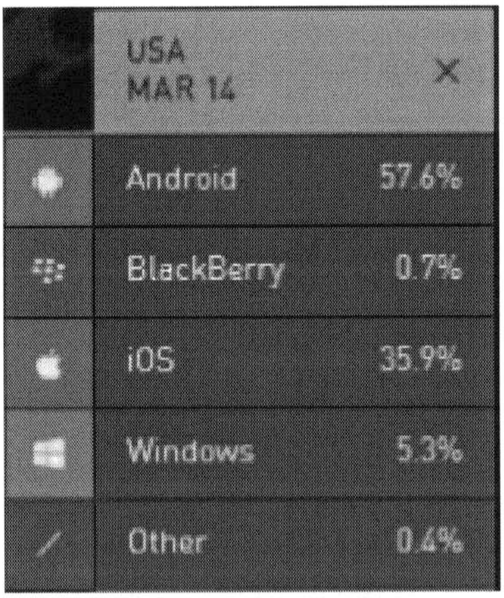

USA MAR 14	X
Android	57.6%
BlackBerry	0.7%
iOS	35.9%
Windows	5.3%
Other	0.4%

FIGURE 5.4 Android is the top mobile OS, with iOS in second. Windows has increased compared to last quarter, and BlackBerry continues to fall.

above, you must seize the opportunity to develop multiple platforms, if the option is available. In cases where there are only enough resources to develop on one platform, the market share that the particular platform controls is very important.

Essentially, the market share of the platform you choose to develop on will be who you will be selling your product to. And to reach the most users out there, you have to choose the market to which most of your users belong. You also have to ensure that the specific platform will continue to command or expand its hold on the market share in the future.

When Apple first came out with the iPhone, it took hold of the mobile device market and reigned supreme for a few years. Shortly after Apple released the iPhone, there was the dawn of age of applications. Businesses saw that more and more people were using their phone, so businesses and developers around the world made fortunes developing applications on the iOS.

A few years after the iPhone, the Android products began to enter the market. These products quickly surged past the iPhone to the top of the consumers' lists. With the emergence of Android, the iOS market began to see a decline, and people who had developed applications on the iOS began to see a loss in the use of their applications, as many users converted from iOS-based phones to Android-based phones.

The same thing happened with Blackberry, when about 2 years ago it was revealed that RIM was having problems within the company. In the example above, many people began to convert from the Blackberry to more advanced and user-friendly mobile devices. In the example used above, if you had decided to invest in the Blackberry at this time, you would have seen your market share decline with the decline of RIM. Figure 5.5 illustrates the margins of phone vendors, and there are 13 vendors who merged, were liquidated, or acquired. None were able to recover.[*]

To future-proof the application you are designing, you must thoroughly understand the history of the market for the platform of your choice. These technology markets can rise and decline very quickly, depending on the next best idea. In order to build an application that is meant to last, you must choose to build it with lasting technologies.

---

[*] Horace Dediu, "Unforgiven: The Consequences of Profit Failure in Mobile Phones," Asymco, April 24, 2012, http://www.asymco.com/2012/04/24/unforgiven-the-consequences-of-profit-failure-in-mobile-phones (accessed June 1, 2014).

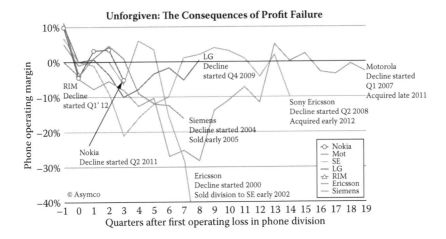

FIGURE 5.5 The phone vendors who merged, were liquidated, or acquired were never able to recover. (From Dediu, H., "Unforgiven: The Consequences of Profit Failure in Mobile Phones," Asymco, April 24, 2012, http://www.asymco. com/2012/04/24/unforgiven-the-consequences-of-profit-failure-in-mobile-phones (accessed June 1, 2014).)

5.4 SUMMARY

Android and iPhone offer two of the leading platforms today, and each one possesses its own unique characteristics and advantages. Android's lead in market share means the potential for app developers to reach a larger audience, but iPhone makes up for this by boasting an attractive audience to marketers that tends to reflect more highly engaged mobile media users with higher incomes. Their strong platform loyalty also bodes well for their ability to retain and expand market share over the long term. Comprehending and quantifying the differences between the users of each platform can help developers make smarter decisions about which audiences and platforms to focus on in order to attract users, drive engagement, produce sales, and to ultimately attain strategic business objectives.

Push Messaging

Lauren Collins

CONTENTS

Push notifications are a tool that enables apps to update users as new content arrives. Users have the ability to turn push notifications on and off per app, or to control the type of notification (sound, on screen, or badges). Instant messages and simple mail transfer protocol (SMTP) email are also examples of push systems. Stock market data, auctions, monitoring, sensor networking, and online gaming are all updated via push-enabled apps as well.

Push services are ultimately derived from information and preferences that are expressed or selected in advance. Developers have been working on this type of technology, called publish/subscribe, for many years. Senders, or publishers, do not have programming to send their messaging to specific receivers, or subscribers. So, published messages are illustrated as channels, having no comprehension of which or how many subscribers there may be. When a subscriber expresses interest in one or more data sets on a particular channel, he or she will only receive messages that are of interest, without knowledge of which or how many publishers there are.

In this example, we will subscribe to channels water and balloon. So, the client will issue a SUBSCRIBE providing the channel names:

```
SUBSCRIBE water balloon
```

All messages sent by other clients to these two channels, water and balloon, will be pushed to all the subscribed clients.

A client who is subscribed to one or more commands will not issue commands, even though it can subscribe and unsubscribe to and from any other channels. The replies of the SUBSCRIBE and UNSUBSCRIBE operations are sent out in the form of messages. The client will simply read a comprehensible stream of messages where the first element is indicative of the message type.

Pushed messages employ a framework in which a message has an array reply with three elements:

1. Subscribe: Client successfully subscribed to the channel noted as the second element in a reply. The third argument represents the number of channels we are currently subscribed to.

2. Unsubscribe: Client successfully unsubscribed to the channel noted as the second element in a reply. The third argument represents the number of channels we are currently subscribed to. If the last argument is zero, we are no longer subscribing to any of the channel, and the client is able to issue any command, as it is now outside of the pub/sub state.

3. Message: A message is received as a result of a publish command being issued by some other client. The second element is the name of the originating channel, where the third argument is the message payload.

Pub/sub was not developed to interact with the key space on any level, including database numbers.

6.1 APPLE PUSH NOTIFICATIONS

Push notifications can be built to notify you of alerts that are triggered. For IT professionals, this is an easy app to build. It can send push notifications to your iPhone® when one of your server alerts has been triggered. After receiving the alert, an app can be launched to instantly view the details of the server that caused the alert. Apple® Developer Support provides the code documentation for the iOS that is necessary to implement and handle alerts on the device, but only provides a high-level guide for the server side.[*] As a software developer, you must notify Apple Push Notification Service (APNS) to ensure sent messages are then pushed to the phone. The provider connects with APNS through a persistent and secure channel while monitoring incoming data intended for its client applications. As new data for an app arrive, the provider prepares and sends notifications through the channel to APNS, which pushes the notification to the target device. Each device establishes an accredited and encrypted IP connection with the service and receives notifications over this persistent connection. If a notification for an application arrives when that application is not running, the device alerts the user that the application has data waiting for it.

Apple's Push Notification Service has three basic requirements:

1. Server: Apple does not provide any front end from which you can send notification, so you will need to have your own server that connects to Apple's servers and sends pushes. As an alternate option, several third-party services offer to connect you to the APNS. They will take care of setting up the Secure Socket Layer (SSL) connection to the APNS and will expose the API you can use to conveniently send pushes. The best services offer a web front end so notifications can be easily sent without writing any code.

2. SSL Certificate: This certificate will identify your app to the service, and is also used to establish a secure connection between your server and the APNS. These certificates are generated free of charge within the iOS Provisioning Portal.

---

[*] "Local and Push Notification Programming Guide," Apple Push Notification Service, n.d., https://developer.apple.com/library/ios/documentation/NetworkingInternet/Conceptual/RemoteNotificationsPG/Chapters/WhatAreRemoteNotif.html#//apple_ref/doc/uid/TP40008194-CH102-SW1 (accessed June 8, 2014).

3. Provisioning profile: This is one of the last steps before your push notification is ready. Once this multiple-step process is completed, you will copy the newly generated .mobileprovision file onto Xcode® or into the iTunes dock. On Apple's side, the provisioning profile will have the certificates from you and Apple, the proper push entitlements, and the `aps` environment key.

For Apple to properly associate pushes to a service with a specific application, APNS does not support any apps with a wildcard bundle identifier (e.g., an identifier with a trailering asterisk such as collins.com.index*).

6.1.1 Register for Push Notifications

Your app must tell the APNS the type of notifications it wishes to receive. Any combination of the following three types can be used in order of the least to most intrusive:

- Badges: Display a small number in a red circle on top of your app icon (shown in Figure 6.1). The badge is generally used to denote the number of unread or otherwise unattended to pieces of content within the app, e.g., unread messages or missed phone calls. An application is responsible for managing the badge number. If the application does not clear the badge, push message(s) will remain in the notification center. After an application receives a push notification, it should remove the icon badge by setting the `applicationBadgeNumber` property of the UIApplication object to 0. Keeping track of the badge

FIGURE 6.1 A badge is one type of push notification in iOS. In this example, the number 2 is alerting the user there are two unprocessed notes in the Drafts app.

value a user sees is imperative since oftentimes the value is relative to another value unknown to the app publisher. Urban Airship offers a great feature, Autobadge, which keeps track of the badge value seen by a user.[*]

- Sounds: Plays the default system sound, or one of your branded sounds. If you choose to use sound, be sure to include it in your app bundle, and it must be in one of the following formats:

  - MA4

  - Linear PCM

  - μLaw

  - aLaw

Push notifications are silent by default. In order to specify an alert sound to be played when the push notification is received, add a sound key to the iOS override section.

```
{
   "audience": {"alias": "user"},
   "notification": {
         "iOS": {
               "alert": "Hear that sound",
               "sound": "default"
         }
   },
   "device_types": ["ios"]
}
```

- Alerts: Displays a UIAlertView with your message, shown in Figure 6.2.

Since alerts force the user to stop what he or she is doing to interact with them, consider how often you will send push notifications. If you abuse your users, they either will turn off the push notifications or delete the app altogether.

---

[*] "IOS Push—Urban Airship Current Documentation," n.d., http://docs.urbanairship.com/connect/ios_push.html (accessed June 8, 2014).

FIGURE 6.2   Push notification in the form of an alert pops up on the user's screen and requires user interaction.

6.1.2 Security Architecture

Apple Push Notification Service exposes certain entry points to enable communications between a provider and a device. Figure 6.3 illustrates the device-facing and provider-facing sides of APNS, both having multiple connections. Multiple providers are making one or more persistent and secure connections with APNS through gateways.

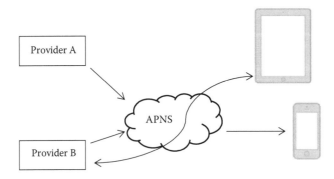

FIGURE 6.3   Multiple providers send notifications through APNS to multiple devices on which their client applications are installed.

To guarantee security, APNS must regulate access to these entry points. There are two levels of trust for providers, devices, and their communications: connection trust and token trust. Connection trust establishes with certainty that the APNS connection is with an authorized provider to whom Apple has agreed to deliver notifications. On the device side, APNS must validate the connection is with a genuine device. Once the trust of entry points has been established, APNS must ensure it conveys notifications only to legitimate entry points. Routing of messages goes through a validation process that is traveling through the transport; only the device that is the intended target of a notification should receive it. Token trust is assurance of accurate message routing, which is made possible through a device token. The device token is an obscure identifier of a device that APNS provides to the device upon initial connection. The device shares the token with its provider. Subsequently, this token will accompany each notification from the provider, which is the basis for establishing trust that the routing of a particular notification is authentic.

6.1.2.1 Service-to-Device Connection Trust

APNS establishes the identity of a connecting device through Transport Layer Security (TLS) peer-to-peer authentication. A device will initiate a TLS connection with APNS, which returns a server certificate. The device validates this certificate and then sends its device certificate to APNS, which also validates the certificate. This process is shown in Figure 6.4.

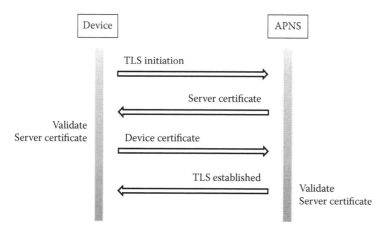

FIGURE 6.4　APNS establishes the identity of a connecting device through TLS peer-to-peer authentication.

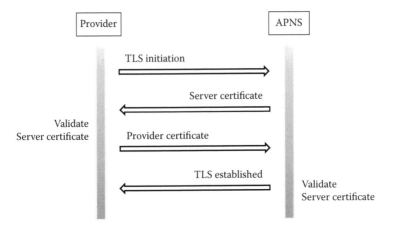

FIGURE 6.5 Connection trust between a provider and APNS is established through TLS peer-to-peer authentication.

6.1.2.2 Provider-to-Service Connection Trust

Connection between a provider and APNS is also established through TLS peer-to-peer authentication. Service-to-device connection trust is similar to the procedure described above. The provider will initiate a TLS connection, obtain the server certificate from APNS, and validate the certificate. Then the provider will send its provider certificate to APNS, which validates it at its end. After this procedure completes, a secure TLS connection has been established; APNS is content that the connection has been made by a legitimate provider. This process is shown in Figure 6.5.

6.1.2.3 Token Creation and Distribution

Applications must register to receive notifications, and this takes place right after an app is installed. The system receives the registration request from the application, connects with APNS, and forwards the request. APNS generates the device token using information contained in the unique device certificate. The device token contains an identifier of the device. Next, it encrypts the device token with a token key and returns it to the device. The process is shown in Figure 6.6.

The device returns the device token to the requesting application as an NSData object. The application must then deliver the device token to its provider in either binary or hexadecimal format. Figure 6.7 illustrates the token generation and dispersal sequence, and also shows the role of the client application in furnishing its provider with the device token.

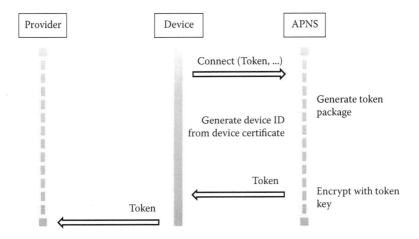

FIGURE 6.6 Applications register to receive notifications, typically right after they're installed on a device.

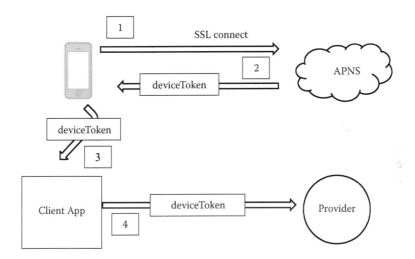

FIGURE 6.7 Sharing the device token.

The form of this phase of token trust ensures that only APNS generates the token, which it will later honor, and can ensure that a token handed to it by a device is the same token that it previously provisioned for that particular device—and only for that device.

6.1.2.4 Token Trust (Notifications)

After the system obtains a device token from APNS, as described above, it must provide APNS with the token every time it connects with it.

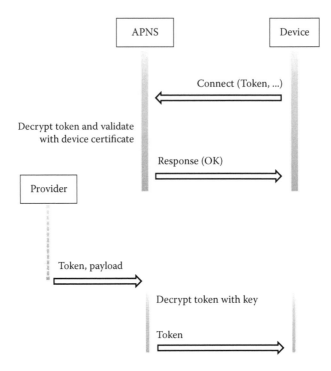

FIGURE 6.8 The flow of a token originates from the device, travels to APNS, and then is sent to the provider once it is decrypted and validated. The provider sends the token back to the originating device, with a payload.

APNS decrypts the device token and validates that the token was generated for the connecting device. To validate, APNS ensures that the device identifier contained in the token matches the device identifier in the device certificate.

Every notification that a provider sends to APNS for delivery to a device must be accompanied by the device token it obtained from an application on that device. APNS decrypts the token using the token key, thereby ensuring that the notification is valid. It then uses the device ID contained in the device token to determine the destination device for the notification. This process is shown in Figure 6.8.

6.1.2.5 Trust Components
To support the security model for APNS, providers and devices must possess certain certificates, certificate authority (CA) certificates, or tokens.

- Provider: Each provider requires a unique provider certificate and private cryptographic key for validating its connection with APNS. This certificate, provisioned by Apple, must identify the particular topic published by the provider; the topic is the bundle ID of the client application. For each notification, the provider must furnish APNS with a device token identifying the target device. The provider may wish to validate the service it is connecting to using the public server certificate provided by the APNS server (optional).

- Device: The system uses the public server certificate passed to it by APNS to authenticate the service that it has connected to. It has a unique private key and certificate that it uses to authenticate itself to the service and establish the TLS connection. It obtains the device certificate and key during device activation and stores them in the keychain. The system also holds its particular device token, which receives the service connection process. Each registered client application is responsible for delivering this token to its content provider.

6.1.3 Notification Payload

Each push notification includes a payload. The payload contains information about how the system should alert the user as well as any custom data you provide. The maximum size allowed for a notification payload is 256 bytes. Apple Push Notification Service refuses any notification that exceeds this limit.

For each notification, compose a JavaScript Object Notification (JSON) dictionary object, as defined by RFC 4627. This dictionary must contain another dictionary identified by the key aps. The aps dictionary, shown in Table 6.1, contains one or more properties that specify the following actions:

- A number to badge the application with

- An alert message to display to the user

- A sound to play

If the target application is not running when the notification arrives, the alert message, sound, or badge value is played or shown. Table 6.2 lists the keys and expected values for the alert dictionary. If the application

TABLE 6.1   Keys and Values of the APS Dictionary

Key	Value Type	Comment
Body	String	The text of the alert message.
Action-loc-key	String or null	If a string is specified, the system displays an alert with two buttons, whose behavior is described in Table 6.1. The string is used as a key to get a localized string in the current localization to use for the right button's title instead of View.
Loc-key	String	A key to an alert message string in a `Localizable.string` file for the current localization (set by the user's language preference). The key string can be formatted with `%@` and `%n$@` specifiers to take the variables specified in `loc-args`.
Loc-args	Array of strings	Variable string values to appear in place of the format specifiers in `loc-key`.
Launch-image	String	The filename of an image file in the application bundle; it may include the extension or be omitted. The image is used as the launch image when users tap the action button or move the action slider. If this property is not specified, the system either uses the previous snapshots, uses the image identified by the `UILaunchImageFile` key in the application's `Info.plist` file, or falls back to the Default.png.

is running, the system delivers the notification to the application delegate as an `NSDictionary` object. Delegation is a pattern in which one program acts on behalf of, or in coordination with, another object. The delegating object keeps a reference to the other object—the delegate—and at the appropriate time sends a message to it. The message informs the delegate of an event that the delegating object is about to handle or has just handled. The delegate may respond to the message by updating the appearance or state of itself or other objects in the application. And in some cases it can return a value that affects how an impending event is handled. The main value of delegation is that it allows you to easily customize the behavior of several objects in one central object.

The dictionary contains the corresponding Cocoa property list objects (plus `NSNull`). The delegating object is typically a framework object, and the delegate is typically a custom controller object. In a managed memory environment, the delegating object maintains a weak reference to its delegate; in a garbage-collected environment, the receiver maintains a strong reference to its delegate. Examples of delegation abound in the Foundation, UIKit, AppKit, and other Cocoa and Cocoa Touch frameworks.

TABLE 6.2   Keys and Values of the Alert Dictionary

Key	Value Type	Comment
Alert	String or dictionary	If this property is included, the system displays a standard event. You may specify a string as the value of an alert or a dictionary as its value. If you specify a string, it becomes the message text of an alert with two buttons: Close and View. If the user taps View, the application is launched. Alternatively, you can specify a dictionary as the value of an alert. See Table 6.1 for descriptions of the keys of this dictionary.
Badge	Number	The number to display as the badge of the application icon. If this property is absent, the badge is not changed. To remove the badge, set the value of this property to 0.
Sound	String	The name of a sound file in the application bundle. The sound in this file is played as an alert. If the sound file does not exist or default is specified as the value, the default alert sound is played. The audio must be in one of the audio data formats that are compatible with system sounds.
Content available	Number	Provide this key with a value of 1 to indicate that new content is available. This is used to support Newsstand apps and background content downloads. Newsstand apps are guaranteed to be able to receive at least one push with this key per 24-hour window.

An example of a delegating object is an instance of the NSWindow class of the AppKit framework. NSWindow declares a protocol, among whose method is windowShouldClose:. When a user clicks the close box in a window, the window object sends windowShouldClose: to its delegate to ask it to confirm the closure of the window. The delegate returns a Boolean value, thereby controlling the behavior of the window object, as shown in Figure 6.9.

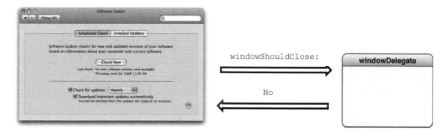

FIGURE 6.9   When a user clicks the close box in a window, the object sends windowShouldClose: to its delegate, asking to confirm the closure of the window. The delegate returns a Boolean value, thereby controlling the behavior of the window object.

If you want the device to display the message text as is in an alert that has both the Close and View buttons, specify a string as the direct value of an alert. *Don't* specify a dictionary as the value of `alert` if the dictionary only has the `body` property.

6.1.4 Localized Formatted Strings

You can display localized alert messages in two ways. The server originating the notification can localize the text; to do this, it must discover the current language preference selected for the device. Otherwise, the client application can store in its bundle the alert message strings translated for each localization it supports. The provider specifies the `loc-key` and `loc-args` properties in the `aps` dictionary of the notification payload. When the device receives the notification (assuming the application is not running), it uses these `aps` dictionary properties to find and format the string localized for the current language, which it then displays to the user.

An application can internationalize resources such as images, sounds, and text for each language that it supports. Internationalization collects the resources and puts them in a subdirectory of the bundle with a two-part name: a language code and an extension of `.lproj` (e.g., `fr.lproj`). Localized strings that are programmatically displayed are put in a file called `Localizable.strings`. Each entry in this file has a key and a localized string value; the string can have format specifiers for the substitution of variable values. When an application asks for a particular resource, such as a localized string, it gets the resource that is localized for the language currently selected by the user. For example, if the preferred language is French, the corresponding string value for an alert message would be fetched from `Localizable.strings` in the `fr.lproj` directory in the application bundle. The application makes this request through the `NSLocalizedString` macro.

This general pattern is also followed when the value of the `action-loc-key` property is a string. This string is a key into the `Localizable.strings` in the localization directory for the currently selected language. iOS uses this key to get the title of the button on the right side of an alert message.

Below is an example where the provider specifies the following dictionary as the value of the alert property:

```
"alert" : {
    "loc-key" : "GAME_PLAY_REQUEST_FORMAT",
```

```
"loc-args" : ["Rita", "Norm"]
}
```

When the device receives the notification, it uses "GAME_PLAY_ REQUEST_FORMAT" as a key to look up the associated string value in the Localizable.strings file in the .lproj directory for the current language. Assuming the current localization has a Localizable. strings entry similar to this,

```
"GAME_PLAY_REQUEST_FORMAT" = "%@ and%@ have invited
you to play Candy Crush";
```

the device displays an alert with the message "Rita and Norm have invited you to play Candy Crush." In addition to the format specifier %@, you can add %n$@ format specifiers for positional substitution of string variables. The *n* is the index (starting with 1) of the array value in loc-args to substitute. There is also the %% specifier for expressing a percentage sign (%). If the entry in the Localizable.strings is as follows,

```
"GAME_PLAY_REQUEST_FORMAT" = "%2$@ and%1$@ have
invited you to play Candy Crush";
```

the device displays an alert with the message "Norm and Rita have invited you to play Candy Crush."

6.2 CODE THE PROJECT IN XCODE

Open Xcode and create a new iPhone project. For this example, we will use a simple View-based Application template, as shown in Figure 6.10, and name the project "Smoke."

An initial step is to inform iOS that our app wants to receive alert notifications. After opening the project SmokeAppDelegate.m (or whatever your app delegate is named), update the application :didFinishLaunching-WithOptions: callback to look like this:

- (BOOL)application (UIApplication*)application:

```
didFinishLaunchingWithOptions:(NSDictionary*)
launchOption
{
//Register for alert notifications
```

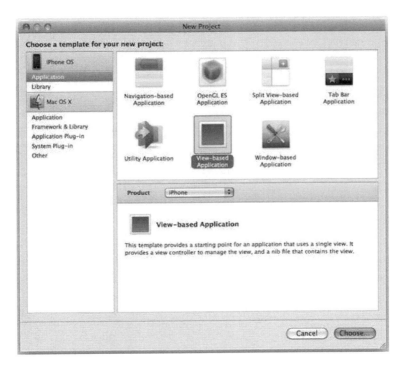

FIGURE 6.10 The selection when creating a new Xcode project is shown. For this example, we will be a using the View-based Application template for iPhone.

```
[application
registerForRemoteNotificationTypes:UIRemoteN
//Add the view controller's view to the window and
display
[window addSubview:viewController.view];
[window makeKeyAndVisible];
return YES;
}
```

Notice we are only passing UIRemoteNotificationTypeAlert, because all we want to receive right now is alerts. If you would like to receive other notifications, go ahead and add them:

```
[...]Types:UIRemoteNotificationTypeAlert | UIRemoteN
```

With these modifications to our app delegate, the OS will begin registering with the Apple Push Notification Service (APNS). If it succeeds, it will send an application.didRegisterForRemoteNotificationsWithDeviceToken:

message to your app delegate with the device token; if it fails, it will send an application:didFailToRegisterForRemoteNotificationsWithError: message. Ideally, you should implement both.

Add the following two methods to your app delegate to handle the two possible outcomes of registration:

- (void)application:(UIApplication*)application:

```
didRegisterForRemoteNotificationsWithDeviceToken:
(NSData*
{
//TODO: Pass the token to our server
NSLog(@"We successfully registered for push
notifications
}
```

- (void)application:(UIApplication*)application:

```
didFailToRegisterForRemoteNotificationsWithError:
(NSError
{
//Inform the user registration failed
NSString* failureMessage = @"There was an error
while trying
to register for push notifications.";
UIAlertView* failureAlert = [[UIAlertView alloc]
initWith
message
delete
cancelButton
otherButton
[failureAlert show];
[failureAlert release];
}
```

Keep in mind that you cannot test a registration on an iPhone simulator, as it does not support push notifications, and it will always fail. You must use an actual device to test your code.

6.2.1 Configure App in iOS Provisioning Portal

Once the basic app is created on our end, it is time to set up Apple's side of things and work in the iOS Provisioning Portal. If you are a member of the iOS Developer Program, log in to the iOS Dev Center and click the link to navigate to the iOS Provisioning Portal. Ensure your account is properly set up with the necessary developer certificates and that your devices have been added. Apple provides a setup assistant for all aspects of this project, if any assistance is needed.

The next step is to create a new App ID. In order to properly associate pushes sent to the service with a specific application, the Apple Push Notification Service does *not* support apps with a wildcard bundle identifier, as previously mentioned in this chapter. So, even though you may have a wildcard identifier set up for development, you will need to create a new one for this app in order to enable push notification support.

As shown in Figure 6.11, click the New App ID button in the top, right corner. Fill in your app's information and click the Submit button.

Locate your newly created, or already existing, App ID in the list and click the Configure link on the right. In the page that comes up, check "Enable for Apple Push Notification Service," shown in Figure 6.12.

If "Enable for Apple Push Notification Service" does not appear, your App ID is probably not configured properly. Go back a step and make sure you did not enter a wildcard bundle identifier.

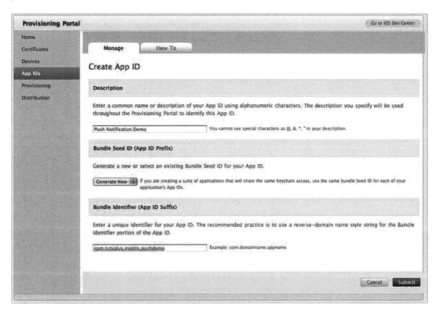

FIGURE 6.11   Create a new App ID to properly associate pushes sent to the service with a specific application.

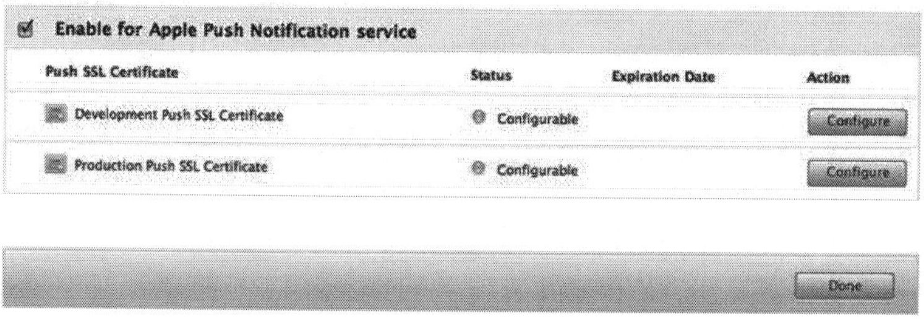

FIGURE 6.12 Ensure "Enable for Apple Push Notification Service" shows up; otherwise, your App ID is probably not configured properly.

You will notice there are two certificates listed in Figure 6.12; one is for development and the other is for production. The APNS has test servers and live production servers. Each uses and requires a different client SSL Certificate. The test servers can be used to send pushes to apps built with development provisioning profiles; the production servers can be used to send pushes to apps built with distribution provisioning profiles (ad hoc, in-house, or App Store builds). Setting up an account with Urban Airship will make sense during that phase.

As we will be working with a development build of our app, click the Configure button for the Development Push SSL Certificate, shown in Figure 6.12. Follow the directions in the box that appears to create a certificate signing request, then click Continue.

Select your newly generated Certificate Signing Request (CSR) by clicking Choose File, and then click Generate to generate your SSL Certificate, as illustrated in Figure 6.13.

Once you see confirmation that the APNS SSL Certificate has been generated, download and install the certificate. Next, we will create a provisioning profile for our new App ID. You can verify a profile has the proper push entitlements by opening it in a text editor. It should have an aps environment key, illustrated in Figure 6.14.

Back in the iOS Provisioning Portal, select Provisioning from the main menu on the left side. Click the Development tab and select the New Profile button. Enter a descriptive name in the Profile Name field and select the certificates for the developers who can use this profile to build apps. If you signed up as an individual, only one certificate will be listed (yours); if you signed up as a company, there may be more than one listed.

FIGURE 6.13   The APNS client SSL Certificate is generated once you select your CSR file.

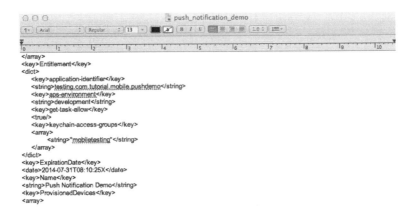

FIGURE 6.14   To verify a profile has proper push entitlements, verify you have an aps environment key.

Choose the App ID you created earlier from the App ID list, and select the devices you want to test your app from the list at the bottom. Then click Submit to generate the new provisioning profile.

After a few minutes, your profile will be issued and a Download button will appear in the Actions column at the right. Refresh your browser if too much time passes without action. Once it does, click it to download your new development provisioning profile.

To install your profile, drag the newly downloaded .mobileprovision file onto Xcode or iTunes in the dock.

6.2.2 Configure App to Use New Provisioning Profile

Navigate to your project in Xcode, and select EditActive Target'Smoke' from the Project menu. Click the Properties tab and enter the bundle identifier you set up in the iOS Provisioning Portal earlier (it was a URL).

Next, switch over to the Build tab. Make sure the configuration selected at the top is Debug. Find the Code Signing section in the table. The second item in the list should be Code Signing Identity, and it should have one child item, Any iOS Device. This setting tells Xcode which provisioning profile to use to sign your app when building for an iOS device. The value for Any iOS Device should be "iPhone Developer (currently matches 'iPhone Developer: [Your name]'[...])."

This currently selected option is the Automatic Profile Selector. It looks at your app's bundle ID (which is what we set in the last step) and finds the provisioning profile with a matching ID (or a wildcard ID that matches). While you can probably get away with leaving this option as is, you can minimize potential sources of code signing errors by selecting the profile to use manually.

6.3 SUMMARY

Using push notifications is a great way to add real-time messaging to your application. iOS push messaging was covered in this chapter, and greatly differs from Android. Android and iOS both require the device to contact their respective push message provider to receive a special string that uniquely identifies a specific app on a specific device. Google Cloud Messaging (GCM), Android's platform, calls it a device ID, while APNS calls it a device token. It is the app's responsibility to receive this identifier from the provider and send it to your application server, so your app can receive push messages from your application server. Both platforms allow for sending messages from one application server to multiple devices and for multiple application servers to send to the same device.

APNS device token is always a 64-byte hex string, and GCM's device ID varies. Push messages sent to GCM always return a reply to the application server informing it whether or not the delivery of the message to GCM was successful. The reply is useful as it can tell the application server whether the device ID has changed or if it has become invalid, and should no longer receive push messages. APNS forces you to intermittently connect to an APNS feedback server that will tell you if any device tokens previously used have become invalid.

Providing a full understanding of implementing cross-platform solutions is important. There are alternate solutions available that provide a unified interface for push notifications servicing iOS and Android. Be sure to check out Urban Airship, PushApps, PushBots, and Amazon AWS Simple Notification Service (SNS).

Mobile Databases

Nate Noonen

CONTENTS

THE BOTTOM LINE does not separate good technologists from great technologists. Margaret Mead said, "Never doubt that a small, thoughtful group of committed citizens can change the world; indeed, it's the only thing that ever has."* At this point in history more than in any other, small and committed groups of people have power. Technology is the great equalizer and does not care if its user comes from humble beginnings or wealth. Unfortunately, the problems solved by technology increasingly cater to those with money and access. There's a place in this world for an easier way to get a cab or to figure out if a restaurant is good. However, there's also a place in the world for technology that serves those less fortunate.

From its roots, technology has been the place for people who are different. As it moves more into popular culture, pressure to make money will be fierce. However, monetization can also mean leaving the world in a better place than we found it. All technologists should aspire to make money and to add value; however, this should not come at the expense of one's humanity. Tim Berners-Lee envisioned a connected society, Richard

---

* See http://www.brainyquote.com/quotes/quotes/m/margaretme100502.html.

Stallman fights for software freedom, and Dennis Ritchie did amazing things under the radar of popular culture. The legacies that these pioneers left should be ones of inclusion. And the very real problems of the world can, and should, be solved by technologists.

7.1 APP DEVELOPER LIFE

When writing the first iteration of software, it is the temptation of every developer to attempt to reinvent some technology that doesn't fit the technical requirements exactly. Often, too many open-source or off-the-shelf technologies are used; this is an unsustainable long-term solution. The best use of a small development team's time is adding features to an application or working to understand and push the limits of an existing framework. The most interesting innovations happen when working within frustrating constraints. A surefire way to fail is to allow a new way of implementing technology that already exists, *simply because a developer doesn't like it*. Just because the control does not do what it was intended to, it isn't time to write one from scratch. Hunker down and see if there are extensible versions online that can be modified to suit the team's very specific needs. Otherwise, it's time for an awkward conversation (or ideas session) with the person responsible for the requirement. If something isn't implemented, there's a good reason for it, and by trying to force it into a specific implementation, the app will most likely be unintuitive to its user base. There is a place for deep customization, but it's when the customers demanding it are giving hundreds of thousands of dollars to support the app, not when they're offering tidbits of feedback.

Development decisions are not made in a vacuum; the decisions are made so that the developer of the application can spend as much time as possible in valuable code and as little time as possible in scaffolding code. Cuts must be made in order to get the first version out. If hallway testing with roommates, colleagues, or significant others reveals something that's frustrating, it's the app's fault, not the fault of the reviewer. Until the people who are hallway testing the app are satisfied, there's no reason to release the app. Until the nicest critics are appeased, the meanest critics should stay away.

7.2 DATA AND API LAYER

One thing holds true of every mobile application: as much functionality, logic, and data as possible need to be behind an application programming interface (API). Even in the case of a data-centric application, data should be behind a web service.

Consider:

1. Every line of code is a possible defect, and as much code as possible should be shared between the various platforms.

2. A well-factored API layer allows for automated testing of functionality, which is difficult, if not impossible, when the functionality is in the user interface (UI) layer.

3. Updates and bug fixes can occur without a client update and its associated turnaround time.

4. Rendering and display of data still need to happen on the device itself, but a data tier designed to use web services should be the goal of every mobile application.

Additionally, there are test cases that can only be executed when the application is in offline mode. Without a separated API layer, these tests are difficult, if not impossible, to execute. This also forces developers to consider what happens when the connection goes down. When faced with intermittent disconnects, team members coming from a thin client- or server-side development role are forced to submit trouble tickets to the network team. In mobile development, intermittent disconnections are the status quo and need to be considered. Figure 7.1 illustrates alternative data caching strategies that developers follow in an effort to steer clear from the troubling offline mode.

While strategies 1 to 3 seem to involve more development hours and higher maintenance, strategy 4 shouldn't be considered. If it seems too good to be true, it probably is.

Another benefit of using an API is that it allows for another path to monetization. Facebook has embraced API, as has Amazon, but for different reasons. The former corroborates steering users to its website, monetizing the users tangentially. Amazon has built a technology infrastructure in an effort to respond to the busiest shopping days of the year. For those nonbusy shopping days, the infrastructure just sits there idle. What they have done is taken the infrastructure and services built internally and exposed them. In turn, web traffic has become a large revenue stream for Amazon. Even if the app ends up not being successful, time

Strategy 1: Full synchronization:

- All data is synchronized on the device during startup
- Edits to the cached data are saved locally
- Edits to the local data are intermittently pushed to the server

Strategy 2: Cached reads, offline writing:

- Increased risk for data breach
- Edits to the cached data are saved locally
- Edits to the local data are intermittently pushed to the server
- Data is cached as it is accessed

Strategy 3: Cached reads, online writes:

- Limited risk for data breach
- Updates performed via web service calls
- No data synchronization required
- Data is cached as it is accessed

Strategy 4: Online read/write:

- Continuous connection needed for real-time updates
- No data synchronization required
- No data is cached, no data stored locally
- No risk for data breach

FIGURE 7.1 Mobile device data caching strategies support the need to develop around offline mode.

should be spent harvesting any IP that is possible. Building an internal app or contributing to a helpful open-source project can turn a failed app into a moderately successful one. The important thing is to gain as much experience and knowledge as possible from the time spent in design and development.

7.3 APP TOOLS

For small development teams, the technologies and platforms should be closely evaluated to determine functionality and usability. The Intel XDK is a HTML5, CSS, and JavaScript development platform that has given universal open access to all developers. This framework allows for a single application to target all the major platforms: iOS, Android™, and Windows®. Additionally, the Intel XDK supports testing across multiple device resolutions. The platform, still in nascent stages, includes a variety of tools and APIs that no application should be without. At this stage, the monetization for Intel, besides being the provider of device hardware, is the tracking mechanism and other metrics. Again, these tools are open to developers for a certain number of downloads. An alternate solution would be PhoneGap, which enables a single HTML5, CSS, and JavaScript, but without the API tracking and add-on features of the Intel XDK.

There are many players in the field for cloud computing. Developers polled from three companies indicated how their work relates to cloud computing over the next year. As shown in Figure 7.2, Microsoft® and Google® have been playing catch up with Amazon® Web Services (AWS). AWS has a cloud computing platform that is extremely simple to use, integrates enterprise features within the public cloud, and offers infrastructure resilience and top performance. At this point, if you are not using AWS, there must be a realistic justification for it. For any public-facing application, using AWS is the way to write software with an elastic scale.

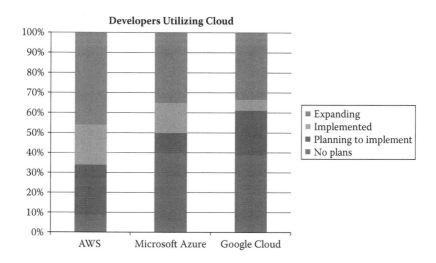

FIGURE 7.2 Amazon Web Services stands apart from other cloud providers.

An application that uses a default private cloud may need to venture into the public space. Azure makes a compelling argument toward the general marketplace, but for a mobile application AWS is a no-brainer.

All developers work within a set of constraints to apply when building code. Their architectural style defines the features used to influence the implementation of software. Representational state transfer (REST) is an elegance that can be applied when building software where clients, user agents, request services and endpoints. REST is one approach to implementing a client-server architectural style, as it specifically builds on top of the client-server model. Another architectural style, Distributed Component Object Model (DCOM), allows a set of logic and interfaces to query services from server objects in a client program. DCOM utilizes HTTP and TCP/IP supporting clients and servers to communicate within the same network.

When Microsoft has fully embraced REST as a first-class citizen, finally backing away from the DCOM-like Windows Communication Foundation (WCF), it's safe to assume that the technology is here to stay. WCF is a framework that builds service-oriented applications; a service endpoint needs to be a continuously available service as a part of Internet information service (IIS) and is hosted within an application. WCF has its place as an internal integration hub. For instance, consider a custom payroll application that needs to talk to a time-off application; however, for a public-facing API, it requires too much configuration. In this example, it is much better to go with a tried-and-true protocol (HTTP) that has been beautifully implemented for web service consumption. One of the most amazing things about being in technology is that the best eureka moments, when explained to peers, elicit the 1990s response of "duh." For example, when a new major software release pushes out REST as a web service, a user (or developer) may show much appreciation and excitement for the noticeably improved differences. Conversely, unaware and unappreciative management may respond, "Isn't that how it has always been done?"

When combining AWS with REST and a desire to eliminate cross-platform code duplication, Service-Oriented Architecture (SOA) becomes the default application design strategy. Service orientation doesn't just mean using web services. Crafting the right service might be the hardest problem to solve; however, as will be discussed later in this chapter, a coarse service that takes in primitive objects can have new functionalities added without breaking backwards compatibility. The best REST services have not altered since their inception because they were designed with as much rigidity as necessary, and with as much flexibility as possible.

7.4 CASE STUDY

Rather than continuing down this path of hypothetical discussion, let's start down the road of designing an actual business application. This discussion will be centered on a business application as opposed to a social- or user content-driven application because the social- and user-driven application space is oversaturated and the world doesn't really need another social media website that initial public offerings (IPOs) for far beyond its actual value.

For this case study we will be discussing an application to manage pickup times for schoolchildren during an emergency. This is in response to a large snowstorm when an early dismissal caused confusion.

The developer of the application was hired by a local school system for the alpha rollout and has the following in place:

1. A cross-platform app enables the school to set up an early release.

2. When an early release is created, a text is sent to a specific phone number when the child is released.

The developer was smart and set up the text-based notification system as a simple Amazon web service that takes in a phone number and a message and routes the message to the phone number. There are now two paths to monetization: the API can be set up as either a service or, because the developer most likely had to use a paid service to send texts, a 99 cent application that enables parents to get emails instead of texts. This both eliminates the cost of the text message service and creates a richer user experience.

The developer was even smarter and, instead of taking in a formatted phone number, took in a 9-digit ID and a string. Behind the API, the string can be either a simple string that is then sent through a notification service or a JavaScript Object Notation (JSON) string that can contain other information.

Having the bulk of the functionality behind an API doesn't limit the possibilities, but depending on the coarseness of the API, can enable more complex and nuanced functionality than a simpler application can.

7.5 UBIQUITOUS VS. TARGETED TECH

There are two types of technology that a mobile application developer can write: ubiquitous and targeted technologies. Ubiquitous technologies utilize components like an object relational mapping (ORM), a message bus, or a text file reader. Many high-level projects have been done numerous times, so it's best to utilize either off-the-shelf software or open-source technologies. They are generic enough to solve the use case at hand.

For this technology, these decisions are inconsequential. The most important aspect to consider is how much time will be spent thinking about the technology, or the amount of effort required tweaking the technology in order to suit business needs. A lot of developers spend countless hours assessing business needs vs. technology, when the real question is whether or not the user base is going to utilize the technology. For example, if the application is only going to end up with 20 users, developers could use a fully normalized schema with NVARCHAR (MAX) columns storing all data and no one would care or notice. If the application works, the technology doesn't matter. Therefore, any shortcut (within reason) should be taken.

Case in point:

- Host the web services in AWS; the infrastructure is robust and readily available.

- Build the UI in PhoneGap or Intel XDK.

- Integrate with OAuth, and make use of every open-source framework; push the app out in front of users across multiple platforms with as little duplicate code as possible.

Targeted technologies are an ORM that adheres to healthcare privacy laws. These targeted technologies work with MySQL, a plug-in framework, for other data technologies. Certain industries such as healthcare, education, and even banking have become stagnant due to the high barrier of entry. One way to remove the barrier of entry is to create a domain-specific framework while solving a specific business problem and making money off the framework and the solution.

Ubiquitous and targeted technologies demand a balance between the hierarchy of business and user needs. Every business unit must work, be fast, and be attractive. If the application can be written and fully functional in 4 weeks, the app will beat any competitor taking 4 months to build. Feedback cycles for mobile apps are much shorter than for commercial software—even if the app is more thorough, possesses a custom ORM, and has beautifully crafted internal architecture. Every application store contains a feedback system that allows an angry user to tear down months of your hard work. The mobile app store is the perfect place for an agile release cycle; the store offers incremental updates, user feedback, and a self-selective audience.

7.6 DEVELOPER VALUE

There was an article a few years ago about a CEO who refused to interview .NET developers for a job. This caused backlash in the development community; however, at the time, the average salary of a .NET developer was x and a Ruby developer was y. The purpose of the CEO's statement could have been about the relative quality of .NET developers; however, it could also have been a statement about how highly developers value themselves. Too often, a developer will leave a well-paid job for technical reasons alone. This, to those of us interested in economics, is tantamount to leaving a job as an artist because the company uses a different kind of oil paint. The structure of consulting agreements, when correctly written, can allow for the harvesting of certain intellectual property by its creator. Figure 7.3 represents the mean income for engineers specializing in certain programming languages.

The reason for all of the above is not to discourage the development of new software, but instead to provide a proper context for its development. Technologists get stuck in the weeds about a specific control or data structure without considering the payoff for their efforts or if their time investment will be worthwhile. The example above simplifies a more complex issue; however, most technologists would agree that the platform for

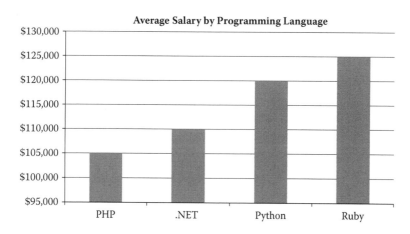

FIGURE 7.3 Database specialists are earning the highest income on average.

development is less important than the problems being solved. In addition, there's a vast amount of knowledge that needs to be gained outside of regular work hours. Anyone who wishes to work in the technology industry, and outlast existing technology, must be constantly learning.

That said, most small development shops making forays into new technologies, such as mobile, would benefit from taking advantage of the knowledge and intelligence of those who have experience in past eras of technology. Objective C is a different animal than Java, so any technological expertise the development team can leverage should be used. Since the economy is still struggling, hiring, training, and paying a solid Java developer with Objective C experience is more feasible for the long term.

7.7 MONETIZATION

The purpose of writing any application is to add value. Whether it is a complex business application or a simple social application, it must be useful to someone. The first thing that must be taken into account is a path to monetization. This is something that is at the forefront of the developer's mind in certain areas more often than others (as mentioned in Michael Carney's profile of Chicago*). There are several ways to make money off of mobile software. And each developer should pick one as a starting point.

The first option is to write an application that is for sale on a market, as shown in Table 7.1. This can create a lot of money quickly. On the other hand, market saturation occurs quickly, and each user is only worth the cost of the application. Heavy or demanding users may use extraordinary amounts of resources while paying the same as light users. In addition, there's no guarantee that anyone besides family will download the app, and hours may be spent building something that very few customers will use.

TABLE 7.1   Pros and Cons of Selling an App on a Market

Pros	Cons
Lots of money if adoption is high	High risk
Choose the right space to market, and may have a better chance of being noticed	Adoption curve may be slow
Multiple modalities are supported and users can get content through all variations	User experience can make or break transaction

---

* See http://pando.com/2013/08/05/chicagos-new-class-of-anti-groupons/.

TABLE 7.2   Pros and Cons of Making Money via Advertising

Pros	Cons
Ability to integrate with any preferred ad network, proving its versatility	Possibly too advanced for green users
Detailed analytics dashboard provides insights about the content performance	Ads tend to appear on sites aimed at a specific demographic that may not match your site

The second option, and the one favored by Silicon Valley types, is to make money via advertising space. This means that heavy users will generate more money; however, people really dislike ads and will use ad-blocking software that stops the revenue stream. Table 7.2 lists the pros and cons of selling an app on a market. In addition, ads must find a balance or risk driving people away as the app becomes more and more saturated with ads as the company attempts to make more money from each user.

Once the alpha release is out, and money is starting to come in, the fun begins. If the adoption rate is as for most applications, it will be slow. However, that doesn't mean the app has failed. As long as some feedback is coming in, and the app is continuing to be improved, adoption will be gradual but may pick up at any point.

At this point, customer engagement is key. If possible, talk with people who have downloaded the app. A personal email should be sent for every download, asking for any feedback and thanking them for spending their hard-earned money or time using your app. The success or failure of your app can depend on the first hundred downloads, and any value that can be gained from user engagement should be utilized.

CHAPTER 8

Mobile Application Development

Scott R. Ellis

CONTENTS

8.1 GETTING STARTED

So far, this book has provided information about the networks, the infrastructure, the technologies used, and their methodologies. In this chapter, for the purpose of simplicity and convenience, we have chosen to develop using the Agile software development methodology and using iPhone® as the platform. iPhone was chosen for one reason: the primary user of this application, the person who will be testing it, already has an iPhone. The objective of this chapter is to take a holistic, 10,000-foot view of building an app from concept to build and through to future development and modeling. This will include a short lecture on how to organize your thoughts and development efforts using a development method that will allow you to focus on the following items:

- The audience for your application, because they are the reason you are building it. If you have the necessity, and the skill, building an app *just for you* is perfectly OK, and don't ever let anyone tell you differently. Everyone thought Marconi was nuts, too.*

---

* It is rumored that after allegedly writing to the Italian Minister of Posts and Telegraphs about his wireless telegraph, Guglielmo Marconi never received a response, and it is rumored that the minister wrote "to [insane asylum]" on the letter. Further, a Groucho Marx comedy routine with Chico begins with: "They said Edison was mad, they said Einstein was mad, they said Marconi was mad…."

- Who they are, really—roles within your audience.

- What features will you include?

- Building your app—*starting simple.*

- Release cycles and publishing your app.

It is critical that you build your app according to a framework. If you have no endgame in mind when you begin, then your path will be very fragmented and disjointed. The lack of focus will present itself in your app, and short of some revolutionary design, you will get few returning customers.

This chapter will include some brief discussion on developing in Xcode. It will conclude with an invitation to become part of an online community where this project is being actively developed.

8.2 DESIGNING THE APP

8.2.1 Defining Need vs. Demand

Generally speaking, demand for a product or service drives design. Apps are no different, but sometimes people don't realize that they want something until they see it. No demand for the successful Twitter app existed until someone saw that the app would fill a void. So, it is a consumer-based need that drives the creation of things. And the demand for a thing becomes measurable only after it has been created and a need has been fulfilled. Meeting a demand where one demand has not been previously met is one avenue of application development. And meeting a need is another. The two are very different things. For example, creating a supply chain for some raw material where the supply chain did not previously exist allows a manufacturer to meet the demand for finished goods. As another example, retail outlets look for underserviced markets to increase their distribution points and capitalize on market surface area. This is meeting demand. Ever since the concept of the department store emerged, no retailer has been satisfying need. They have merely been finding new and innovative ways to meet demand.

Need, then, represents the fulfillment of a requirement that is satisfied when the demand for it is met. The Wright brothers fulfilled a need. United Airlines fulfills demand.

8.3 PERFORMING AND REVIEWING CASE STUDIES

Those who are unaware of history are destined to repeat it.

—GEORGE SANTAYANA

Case studies provide vantage points into what we are about to do, and they are the examples of those that have gone before us and have previously done it. The best case studies are built on your own past experience. Based upon them, by carefully reviewing the things that we have done in the past, and the things that others have done, we can hopefully prepare a more thorough and navigable path forward through development.

8.3.1 Case Study: Relativity

In the early part of the aught, the American justice system had no real regulations about how courtroom *electronic* discovery requests should or would be met. A lot of different software vendors offered products that would allow litigants to gather paper and digital files and produce, or disclose them, to each other. *Blowbacks*, an industry term for printouts of discovery, were common and the printing and paper industries were in hog heaven.

The software companies had been designing their software to meet the demands of their clients. In December 2006, there were several major changes to the Federal Rules of Civil Procedure (FRCP). The changes were not at all subtle. They were deep, and affected a number of areas, and they severely impacted the software industry that had been providing discovery services to litigants. Primarily, the new rules made it acceptable (authorized) for opposing sides to request native documents in their entirety. By 2006, the industry shift away from paper was in full swing— things just weren't printed anymore. But the courts were still allowing litigants to force each other to deliver only paper discovery. This was expensive, and caused a lot of unnecessary work. Enter the new rules of civil procedure: everything could be kept digital. Nothing had to be printed anymore.

Thus there was a need: litigants had a need for an affordable software platform that would allow them to seamlessly review thousands upon thousands of electronically store documents. There were very few products that could meet this need, and the ones that could do so were poorly designed. Everyone was a paper handling shop, and they lacked the agility to swiftly tool their platforms. So a need arose for a product that could do this.

You can see then that *need* is an unfulfilled demand that may be known or unknown. Demand is just one-half of the supply and demand equation—demand is generated by a known element. The two things should not be confused. Need is the thing that should drive your app development. It is what will lead you to create the next killer app. Need can also be thought of as something that is generated when an existing industry cannot, will not, or is unable to meet the demands of customers. Sometimes, a need is met by a product, and the demand is high both for that product and for an easier way to do it, such as the loom, which fulfilled a need to simplify fabric production, and thereby met an observed demand. Being able to invent a way to mass-produce something that had previously been incredibly expensive is another factor to consider when inventing.

8.3.2 Case Study: One-Payment Methods

One thing will always be true, and this is that a thing should be easy to use. If it is not easy to use, then no matter how cool it is, it will find no traction. For example, in 2012 I began to see a new payment option appearing at some of the businesses I frequent. I was intrigued by the idea of converting my smart phone into a wallet. I have more credit cards than I can possibly carry, and I thought that the ability to have them all in one secure app on my phone would be great. I attempted to set up an account with one of the first "easy pay" apps that I came across. Unfortunately, the sign-up process lacked usability, and so I discarded it. It boasted a 30-second sign-up time, but after 10 minutes I called it quits. Nothing will lose a customer faster than making false and misleading claims about an application. The placard had made a claim that signing up was fast and easy. Fast-forward to late 2013, and a local convenience store had a tablet/kiosk setup at the counter. The placard says, "Grab a blue card and scan it."

And so, now I have a new payment option, and hopefully it will spread and I can stop carrying a fat wallet. Eventually, the world will be paperless as well as wallet-less.

8.3.3 Case Study: Photos

In another example, more relevant to this chapter, a friend of mine complained to me that our county's forest preserve district should provide some sort of service that allows people to view photos that have been taken within the forest preserve via an app. Currently, many of the forest preserve location websites do allow for posting of photos, but they are generally by county, are not that specific, and don't have an app that is simple to use. I am not sure of his exact motivation, that is, what his use case/story is, for wanting this, but the idea intrigued me. Some similar applications exist, but they required a small purchase. Larger applications, such as Flickr, allow geotag searching, but the interface and the entire idea is not focused on geotagged images. I began to think about how such an application would work, and what its purpose would be.

Ultimately, a software application can be thought of, at least in abstract terms, as an intersection between what can be stored and managed digitally, and what occurs in the real world. The lack of a good intersection point that allows for ease of use and management creates, quite literally, the need. The amount of people that have the need creates the demand. Just because something is needed does not mean it will be in demand. For the purposes of this chapter, the need is to build a simple application and explain how it is done. We will do this using the Agile development method, which is a software development framework with which the authors are familiar, and is one that is especially useful to mobile app development because it stresses simplicity.

This chapter expands forward from this case study, and explores the feasibility of creating an application to meet the need suggested by my friend.

8.4 AGILE DEVELOPMENT

I am not an expert on Agile, but working as I do at a software company that uses this methodology, I have been through training and have attended many sprint release reviews, planning, and retrospectives. My team uses scrum to stay on top of problems that are escalated to us, and we use a daily stand-up to meet with our peers and discuss any challenges or roadblocks to solving the problems. I considered putting *stand-up* in quotes, but then I realized that we actually do it standing up. Many scrum masters will actually ask you to stand up during a stand-up. They say that it helps improve circulation and thinking, and brings greater attention to the process.

I have enough knowledge to put into practice the methods of Agile software development, and provide an explanation of sufficient depth that will suffice for the purposes of this chapter. I have worked in software development for most of my adult life, with a few breaks here and there to do other things, such as fix airplanes and rebuild tractor engines in a barn in central Illinois (which was a rather short-lived career—barns get really hot in the summertime).

Agile, in a nutshell, focuses on building very simple features that focus directly on user needs, which are expressed through the stories the users, themselves, tell. By starting with the simplest of features, assembling them into a working model, and then presenting them to the stakeholders—the end users or those who know the end users the best—early, one can make corrections to the problems, release the features, and then focus on the next iteration: more features. Generally, a development cycle in Agile is called a sprint, points are assigned to user stories based on perceived complexity, and each sprint lasts about 2 weeks.

Agile has several roles and terms that must be defined. At any time, this author may inadvertently or intentionally refer to one of these items. The roles and terms that are essential to understand Agile are as follows:

Stakeholders: These are the people that drive product development. They represent the end users of the software application.

Product owner: This is the person who drives the need and defines the demand created by stakeholders.

Product manager: Assists in defining objectives and creating and managing the workload.

Scrum master: Tracks and manages the burn rate of the team.

Developer: Builds the application. Makes lots of mistakes and calls them features.

User stories: These are the tales that the people who use your app will tell you about what you should do to improve your app. They are the rationale behind feature requests. We will develop user stories for this app in Section 8.5.

Roles: Different users of your application may have different roles within the app. The bigger the app, the more roles. The different roles of this application will be discovered and explored in the following section.

Research: Finally, let's do some research. There are many applications that perform logging, and perform post-shoot geotagging. The question boils down to one of a value-add proposal. Where do we find a niche? And how do we serve that niche? What is the simplest, first step that we can take toward attaining the goal of a more robust application that can meet a very real need? Does this application already exist? Can this application be built on an application program interface (API) to something such as Flickr, and then allow people to view photos that have been geotagged in their location?

Most importantly, what do the people who will use the application want to see in it? For this, we develop user stories.

8.5 USER STORIES

User stories in Agile are little nutshell-sized vignettes that explain the purpose for interstices between real life and data points. A user story explains what a user wants to do in very strict terms of the user's life. It objectifies and parameterizes the user's real-world, task-based objectives, one at a time.

Let's iterate through some steps and try to develop a user story. First, we ask a few questions:

Product Manager (PM): What do you need?
User: As a photographer, I want a way to differentiate the pictures I take from what everyone else is doing.

This is good, but where does it really take us? It has no explicit touch point with the electronic world, and it provides no real direction. It is not a user story, it is wishful thinking. So we ask the questions again:

PM: What do you need in order to accomplish that? Be more specific.
Photographer: As a photographer, I want to easily be able to organize and share my photos.

OK, they have tools for that. What is wrong with the tools that exist? What type of photography do you do? Notice that each iteration of the user story raises additional questions.

Photographer: As a nature photographer, I want to easily be able to organize and share my photos.

OK, now it starts to make sense. Who is your audience? Who would look at these photos? Who cares?

Photographer: As a nature photographer, I want to easily be able to share my photos with people who are interested in going to the places where I have taken photos, or who are already there, or who may have been there while I was there.

Visitor: As a visitor to an area, I want to see photos that have been taken here that are relevant to what I am doing (birding). I also see that there are a few photographers in our group, and I would like to see photos that they took during our outing.

We now have an intersection point between the photographer, the digital medium, and her audience. The Internet acts as the forum. The photographer wants to be able to take photos, track where she has been as she takes them, and then share them in a way that will represent where she has been and will allow people to see where they are, and only see photos of that location. That seems easy enough. Right? She takes photos with her iPhone, they get geotagged, and the information we need is already there, right?

It is important that you ask probing questions of users. Had we proceeded and built an app that did not consider all angles, our end user would have responded with, "This is useless to me." As it happens, most of the current generation of digital single-lens reflex (DSLR) cameras in the hands of consumers and professionals today *do not* come equipped with a GPS. A GPS device can be purchased that can tag photos, but such devices are not available for all cameras, and some will only perform logging. A logger and its accompanying software can then be used to identify the location of photos based on the timestamp of the photos. All photos are timestamped using the time from the on-board camera clock, and the logger software can use this information later to tag the EXIF data. Now, our requirements gathering is nearly complete, and we can begin to think about building our app.

8.5.1 Actors

Now, some background: I have a vocation in life, and it is photography. In particular, bird photography. My son also loves birds, and it is his passion that escalated my interest in photography to the level that it is at now. This sets the stage for the *type* of photography that this application will service.

In Agile, every application that gets built has, behind it, "actors," and so the development organization process begins with the actors, descriptions of them, and their goals. The descriptions of the actors I have created include a short vignette to describe their use:

The strategist: This is someone who doesn't necessarily have a vested interest in actually using the app, but has an interest in seeing it succeed. One such strategist will be the parks. The park wants to see an application that is easy to use, free of ads, and showcases the wildlife at its location.

The scientist: The scientist wants an app that she can use to research. She wants to see bird counts, time of day, location, health of bird, etc. She has an interest in high-quality photographs, but seeing many photographs of the same bird at different angles may be appealing. She wants to be able to view birds at any location, from any location. *Only* users with an .edu address will be able to do this.

The birder/visitor: The birder has many of the same desires as the scientist, though he may be more interested in seeing high-quality images, and having the ability to see images from trips that he took that verify his sightings. Photographers and birders generally intersect in their habits.

The photographer: The photographer will be the principal content provider. As such, most photographers will not be able to be moderators, too.

The moderator: One possible way to promote people to moderators is that after a predetermined number of bird views, anyone who is not a photographer will begin to see a "thumbs up" button. This will enable moderation by voting, and more helpful and better photos will be presented.

8.6 NAMING YOUR APP

So let's build a mobile app. There are many birding applications in the world. Our goal is to build something unique, and so fundamental and simple to use that it will be memorable. A couple of key considerations when naming an application come to mind, and here are some simple requirements for name selection:

1. Easy to spell.

2. Relate to the product.

3. Easy to remember.

4. Unique meaning.

5. Should not be overly confining—as applications grow in functionality over time. It may come to be used for purposes beyond your original intent.

Now, let's take a look at the definition of a particular word I like, and see if it covers these points, and meets the above five requirements.

- car·di·nal

  ˈkärd-nl,ˈkärdn-əl/

- noun

  1. A songbird of the bunting family, found in the New World. This bird has a stout bill and usually has a prominent crest. The male is brighter and more solidly red in color than the female.

- adjective

  1. Of the greatest importance; fundamental.

     "several cardinal points must be kept in mind"

     Synonyms: simple, basic, fundamental, primary, crucial, pivotal, prime, key, essential

- mathematics

  1. Cardinal numbers (cardinals) are numbers used to count things, such as one, two, and three. It answers the question of "How many?" Something that lacks cardinality is missing the attribute of being able to be counted. The cardinality of something indicates how many of that thing there are. A cardinality of "one" means that there can be, or is, only one of a thing.

We see that the word *cardinal* has fantastic symmetry in it. It's a word that means bird, important, and also has a mathematical component to it. With the exception of the religious definition (not shown), it has great relevance to the product.

1. Easy to spell: *Cardinal* is both easy to spell and a common bird that everyone knows. More importantly, this will be the first bird that people will see.

2. Relate to the product: This is an application about birds, and to some extent, bird identification. More on this later, but the iterative nature of counting (cardinality) and the iterative nature of phylogeny (identification) pair well together.

3. Easy to remember: Since many of the initial users of the product also know ee (my son), they should have no trouble appending *cardinal* to his initials.

4. Unique meaning: Cardinal is a nice name, but in order to help it along in this area, I'm going to prefix it with my son's initials, *ee*, to make *eeCardinal*. My son wants to be an ornithologist some day, and as such I imagine he will someday become one of the chief evangelists of the product. He is also the chief bird counter and photo identifier of the project. He is developing the system of tagging that will allow the project to work.

5. Should not be overly confining: *eeCardinal* suggests something about birding, but it does not pin the application to just a photography by location app, or identification only, or just something about cardinals. The name will hopefully, to the scientist, suggest a more significant approach to identification than what might be inherent in the standard, book-like apps that currently exist. It will also allow the app room to grow, and mature over time, and develop in a manner that serves its users.

8.7 PLANNING TO BUILD

Typically, an application would be planned out, with wire-frames, storyboards, etc. This application has one simple goal—to create the simplest app possible. We want someone to be able to "start up" the app in logging mode, shoot some photos, and then be able to tie the photos back to a particular place. The photographer wants to be able to do this, post a load of photos to Flickr, and she wants to be able to access the app in "visitor" mode and see her photos turn up. The app will not store photos, nor will it allow a registered user to view photos under visitor mode. It will, essentially, be a highly targeted search engine.

8.7.1 Location

The location is the intersection of the real world and the virtual one. In abstract, software terms, the location then becomes a portal to the works of photographers, and it is one that only works within the proximity of the GPS location of a photo. In addition to geotagging the photos with a special coordinates tag in Flickr, the application will also create a tag with the location where the photographer "checked in" that it will associate with the GPS log. Additionally, some cameras are equipped with geotagging devices. Once the check-in at a location has been performed, it initiates the location processes. The user will stay checked in until the next check-in, or a check-out. If the photographer forgets he is checked in, photos not taken at the location may be inappropriately tagged. There may be some "sensitivity" settings that can be set. For example, if a photographer sets the radius to 10 miles, all photos taken within that radius and during the interval will be tagged with the location. Future versions may have a lookup and location search capability for scientists only, but for now the sole purpose of this application is to show the photos to the location visitors, and to show only picture of birds.

8.7.2 Getting Set Up

iPhones are very common among birders due to the prevalence of birder-specific apps. This fact, coupled with the phone in use by the photographer, means that we will be building an iOS app. This chapter, while geared toward beginners, is also geared toward beginners who have a technical background. Most developers of iOS apps write in a language called Objective C. Here is a list of words to review, and if the words are familiar to you, then you should be just fine:

- Polymorphism
- Inheritance
- Class
- Function
- Method
- Call

If these words are near to gibberish to you, there are many very good tutorials that you may want to work through before digging in and building an app of your own. W3 Schools offers some very nice tutorials on

Javascript, and there are some very good free or nearly free trials on Cocoa®
that are available on the Internet. A Google search for "how to build an
iOS app" will set you on the right path, but don't feel like you have to pay
$100 for a course on this stuff. There are many examples and low-cost
tutorials available. Sams iOS 7 in 24 hours is a great tool, and is the one
that this author worked through before development of this application.

Programming is not difficult. The discipline is one of understanding
syntax and order of operations. If you are able to converse deeply, and
exchange ideas with other humans in a meaningful way, if you've ever
written a large term paper, and if complex tasks like wrought-iron puz-
zles and locked room mysteries engage you, then you can probably program and write software. Other skills that are similar to writing software
include drawing a picture, giving directions to someone to perform some
complex task, and cooking a large holiday meal.

If you've mastered Objective C, but are new to Mac, the GCC compiler
comes built in to the platform, and you can build and compile files from
the terminal window. There is no shortcut key to open terminal, but press-
ing Command + Space will open a run line similar to the Windows® + r
command that you can use to open a CMD window. Since time is always
of the essence, here are some convenient steps that you can follow to create
a shortcut key to launch the terminal window:

1. Launch Automator and click [New Document]. Choose the gear icon
   for service as the type, and click [Choose]. In the [Service Receives]
   drop down, select No Input. From the Actions Library, click on Utilities
   and then drag the Launch Application icon to the pane on the right.

2. Select the Terminal application in the drop-down list of Applications
   by clicking Other and then typing in "Terminal." Highlight it, and
   click [Choose]. At the top of the Automator window is a subdued button that allows you to name your service. Call it "LaunchTerminal."
   Close out of the window. Use Spotlight to find LaunchTerminal.
   Click on it, and then click the Install button when it appears.

3. Now that it is running as a service, you can assign a keyboard shortcut to it in system preferences:

   System Preferences ⇒ Keyboard ⇒ Keyboard Shortcuts ⇒ Services

Your new shortcut combination will work in most applications. It
doesn't, for example, work in Notes. You'll need to click out to one side

to get it to work. It does work when tapped while editing this chapter in Chrome.

8.7.2.1 Tools Needed

Many developers have Windows-based laptops, and they also have an iPad. The underlying framework used to develop software on iOS and OS X, Cocoa and Cocoa Touch, brings together a collection of very well designed object-oriented frameworks. Cocoa Touch isn't a tool that you can use or see; rather, it is like unto a standard, or generation, of a collection of functional building block classes that includes storage elements, interface items, and other useful tools.

Xcode, then, is a reference framework software that allows the developer to assemble and compile applications. It is an integrated development environment (IDE) that allows you to build and develop the graphical user interface (GUI or UI).

Unfortunately, you can't use Xcode on Windows. There are programs such as Marmalade and Monotouch that can be used to develop cross-platform applications. Marmalade allows for a single, centralized code base that can be compiled for multiple platforms.

- iPhone/iPad—testing platforms

- Mac Computer—development platform

- Xcode—development environment (GUI interface for writing code)

On the Mac OS X, you can also compile the code using a terminal window. One *could* write the platform on a Windows machine and compile and test it there, but ultimately it would have to be ported to a Mac so that it would run there, too. Compiled code won't typically run on an OS other than the one where it was compiled.

8.8 BUILDING IN COCOA WITH XCODE

Cocoa uses a paradigm of development called the model view controller (MVC). The following sections explain each aspect of the MVC.

8.8.1 Model

Xcode is object oriented. The model of Xcode is its ability to allow you to group application functions into objects. It allows you to create a micro-universe that relates to your real world.

8.8.2 Views

Views present the model data, and allow it to be edited. A view is really just that—a view. It takes the data from the controller and presents them to the user. If you happened to be reading this book on an iOS device, you are looking at a view right now. Think of a view as being a bit like a web page for your iOS, and you are spot on. They are the only way that users have to interact with your data—a view *is* the graphical user interface in an iOS app. Much of what you read about views is heavily jargonized. At the end of the day, the paradigm exists already in the form of frames, sub-frames, parent frames, etc. Thinking about them in terms of frames may be helpful if you have a strong background in web development. If you are a Structured Query Language (SQL) nerd, then think of views as being similar to SQL views.

8.8.3 Controller

Controllers negotiate logical operations between views and models. They house the structure and act as the repository and go-to source for all code behind. They *control* the views, and every view has to have a view controller that houses its functions.

This philosophical approach to the organization of structure and form provides a simplified approach to application development. So simple, in fact, that even this author could figure it out! Whatever application you decide to build will be easy to architect, deploy, and when you need to come back and make changes, it should be straightforward and easy.

The first thing you will need to do is download Xcode. But be fore-warned: as of February 2014, it tips the scales at 5.23 GB. So, it might take a couple of hours to download.

The other consequence of using the MVC designer is that you can alleg-edly build an application that requires no code whatsoever. This would be a very simple project, and would be the moral equivalent of a set of static web pages. Your concentration is on the architecture of the application. Glue code is eliminated by the Mac Cocoa bindings. You piece together your application using the GUI quickly, and go to market sooner.

8.9 APPLICATION DESIGN AND EECARDINAL

The first step in the designing of an iOS application involves possi-bly one of the more difficult choices the developer/designer must make. What format of interface design do I want? For this application, which will be called eeCardinal, the architect created a basic Visio diagram that sketched out the basic workflow and design of the tool. This diagram will map the characters and their activities as entities, and overlay them on a process flow diagram that can then be converted into layouts and views in a software program.

We will get to the whiteboarding process, but first let's discuss a little more about who the application will support. For the purposes of this application, focus and emphasis are placed on two roles:

- The photographer: The photographer will have the option to set up a new account. She will also have access to the geologging feature and a check-in function.

- The visitor: Visitors will only be able to view photos in the location where they are. Photos will not, in this iteration, be search capable, and you will not be able to navigate to a location. An arrow may appear on the phone that will point them in the direction where the most images appear.

The process diagram in Figure 8.1 lays out the events and the interactions that will occur within the framework of eeCardinal. It is also a very neatly defined criteria for success. The successful launch of this application will include all of the features of this diagram.

8.9.1 The Application: A Functional Description of eeCardinal

Now that the process and architecture have been defined, and user stories written, it's time to write an outline that explains everything that has been described up to now. This is a very high level requirements document. It will translate into a project plan and will (hopefully) only take 4 weeks to build a single sprint.

1. eeCardinal will be available in the App Store.

2. Photographers will download it.

3. Birders will download it.

4. There will be no ads.

5. There will be a $1 per year subscription fee, which will help offset the costs of keeping the app on the App Store.

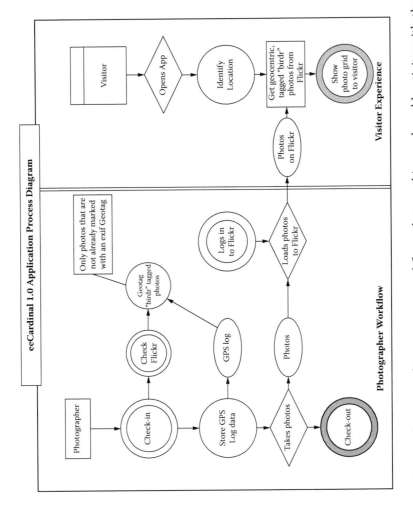

FIGURE 8.1 The two roles are decomposed into process workflows that combine real-world activity with the virtual framework required to support it.

6. Photographers will have to have a Flickr account. This version will offer no alternative method.

7. First-time users of eeCardinal will be presented with two buttons:

   a. Visitor

   b. Photographer

8. There will be a little gear-shaped button in the lower right-hand corner that will allow the user to change this at some point if he likes. Once this choice has been made, it will be remembered by the app.

9. Aside from the visitors' locations, some basic configuration options, and photos that need to be downloaded so that visitors may look at them, eeCardinal will retain no information in the phone or allow it to be sent anywhere other than to be used for the purposes of locating relevant content and showing it to the user.

10. Aside from the photographers' locations, some basic configuration options, and the most recent log that has not yet been synchronized, eeCardinal will retain no information in the phone or allow it to be sent anywhere other than to be used for the purposes of locating relevant content and showing it to the user.

Note: Once the photographer has synchronized successfully, the log will automatically be deleted. She will not be able to "resync" a given log. If she forgets to load some photos, she will need to tag them manually. Some future version may request this feature if it is requested by a photographer user.

11. When a photographer gets past this screen, he will be asked to enter his Flickr information and authenticate to Flickr. These credentials will be stored locally in an encrypted way.

12. The photographer will be asked to agree to terms of use, and she will be asked to fully read and agree with each one, and presented each term on a screen that she must swipe to agree to each. These terms will ask the photographer to agree with the following:

    a. Images that she loads to her Flickr account that are tagged by eeCardinal will be available to all users of eeCardinal, with the only restriction being that the photos can only be viewed while the birder is in the same general location as where the photo was taken. Photos will not be available offline or off-site.

    b. Photos loaded to Flickr will be subject to being tagged with geo-location and park name when they synchronize.

    c. She agrees to have photos moderated, and to not spam the feed.

    d. She agrees to the recommendation of posting no more than three photos of a single bird species per visit.

    e. She agrees to name her photos of birds with the common species name, without misspellings, and to allow the application to add tags to properly identified birds.

13. Once the photographer has completed this, he will see two buttons:

    a. Log

    b. Synchronize

14. The photographer taps the log button, and it goes to a page that asks her to name the location (if it isn't already in the database), and from this point forward, for up to 8 hours, the app will log her GPS coordinates.

15. If the photographer chooses to synchronize, the app will search for any photos on his account that are tagged with the word *eeCardinal* and that have timestamps that match, and it will (if it can) append the GPS coordinates to the EXIF data. Otherwise, it will place the location name into a tag and deal with the GPS coordinates in a different way.

16. If the photographer chooses not to log or synchronize, she can swipe with a three-finger motion to the left, and this will take her to the view screen, which will show thumbnails of the first 12 photo matches found. Swiping to the left will reveal additional screens of birds.

17. The app will always remember that it has a log that needs to be digested.

18. Visitors will not see a log or a synchronize button. They will be taken right to the screen that shows pictures of birds, and they can swipe to view more.

8.9.2 Estimating the Workload

Creating an application is a very creative, stimulating exercise, and many of the concepts and features listed in this chapter will certainly not make it into the first iteration. It is important to document everything, and then prioritize the features and assign points to each. Often, tasks can be broken down by a complexity system that is based on the Fibonacci series (–0, 1, 1, 2, 3, 5, 8, 13, 21, etc.), and the last two numbers in the series are added together to get the next. Some scrum masters will use animals, for example:

1. Mouse

2. Cat

3. Wolf

4. Antelope

5. Buffalo

6. Elephant

As you can imagine, an elephant is a really big user story, whereas a mouse is a task that is clearly something that will be quick and easy to accomplish. From the above information, and future version ideas, a task board can be created where each task has a developer assigned to it, a stakeholder, etc. This becomes the storyboard for the application. Often, this will be something as simple as sticky notes stuck to a wall, or note cards attached to the wall with painter's tape (look for the blue masking tape, but be careful, it can pull the paint off of the wall). Tasks will move across the board as they progress to completion.

8.9.3 Future Versions

After some time, the app will have accumulated a rather large number of photographs. As it matures, it is conceivable that the app could contain references to all of the most common species. The birder also wants to use this app to help identify a bird, and wants an app that is simple and doesn't even require knowing avian anatomy. When in "browsing" mode, which is the standard mode, swiping in any direction will always show you the next bird, and pinching will always show you a 3×9 thumbnail view. In the future, the app could be designed to include access to an "identification mode" (eeMode), which is described here. The premise is that, using a multidimensional matrix, simple hand motions can be used to identify birds. By creating a mathematical model that describes birds along a finite number of vectors, and with subdomains containing additional vector fields, identification of a bird in many cases could be accomplished with as few as three hand motions, and no more than nine. By calculating the vectors as real-valued functions, and by using Newton's method to manage the vectors, more rapid and intuitive decisions can be made. There will of course be exceptions, and special logic will need to be introduced for some species. Primarily, this app will be most appealing to the masses, who are just looking to get a rough idea of the bird they saw:

- To enter eeMode, swipe upwards with three fingers. A screen appears that has nine different birds. The user taps the one that most closely resembles his bird.

- To exit, swipe three down.

- A small icon of a book will appear in the upper right-hand mode as a visual cue, along with a one-time-only alert.

- In eeMode, the following gestures will provide rapid identification of birds:

  - A shake always goes back.

  - Swiping upward means that the bird is taller.

  - Swiping down means it is smaller, or shorter.

  - One finger left means the beak needs to be shorter; two fingers left means longer. This will be intuitive, as most photos loaded will have the beak to the left.

  - One finger to the right means the tail should be shorter; two to the right means it should be longer.

  - Double-tapping means show more of the same, in different scenes. Shake phone to go back to top.

  - Three-finger pinch means a smaller bird.

  - Three-finger spread means a larger bird.

  - Two-finger pinch zooms or shrinks.

  - Tapping and holding on an area of the bird means that the color or feature you are touching is wrong, and to show the next bird that doesn't have that color.

8.9.4 The Whiteboard

It doesn't matter whether you use a real-life whiteboard, something like Visio, or a pencil and a sketchpad for this, or if you decide to do the entire thing as though it were a cartoon. What matters most is that you get a real-life simulation of how your application will work, without ever writing a single line of code.

iPhones seem very simplistic and clever in their architecture—after all, they fit in the palm of your hand, so they must be simple. This is deceptive. An iPhone also has a CPU, RAM, disk, and network considerations. You cannot build an application and avoid thinking about or considering these things. If you do, you will inevitably bite off more than you can chew. Even with its flash-based memory, the architecture of the bus is still such that the tremendous efficiency difference between in-memory processes and external processes (going to the Internet to get a file, trying to manipulate data that won't fit into RAM, etc.) can be the life or death of your application. Follow the 2-second rule: no action should take longer than 2 seconds. The very best application will work under the most heinous of conditions, and it will work well.

So, where do you start? With the stakeholders, and a shell of an idea, you get them in a room and you show them what you are thinking. In Figure 8.1 I introduced the concept of an application process diagram. Now, the application pages must be designed and laid out, and abstracted from the process diagram. You draw it out and explain the advantages, and you get their input. Then you begin to build, and as soon as you have something, anything, that works—even if it is just a little bit—you demo. In Agile development, we call this a sprint review.

- Create a Visio, or some sort of flowchart.

- Before you begin to storyboard your application, it's a good idea to actually hit the whiteboard with it. Get your dry erase marker and get busy sketching out the myriad ways your application could potentially work. I've found that the best approach is to sketch out the entire application, and then break it into pieces. Rework how the pieces function. For example, I recently whiteboarded an application that would allow users to load data to a website and create a daily feed. That is, after the initial load, the data would then be set up to feed up to the system automatically, every day. There needed to be some bare minimums met before the data would be allowed to be posted, and there needed to be a process on the management side of the application to approve or disapprove of a submission.

- Upon completion of the first drawing, I drew a circle around one piece, and then reimagined how it might work, and then asked the stakeholders their opinion, and we discussed the pros and cons of each approach.

Note: This approach truly requires that you let go of your ideas, and that you not become attached to them. Just because something happened in your head and you drew it on a board does not make it golden. *Do not* do this sort of work if you are only looking for an ego boost. People will sense that, and one of two things will happen:

- They will try to please you by telling you it looks great, and you will publish a piece of egotistical garbage.

- People will not take you seriously and will not view your design objectively, you will redesign it to please *their* egos, and then you still publish a piece of ego-driven garbage.

8.10 DEVELOPMENT PLAN

Now that a solid functional description has been created, a plan of attack is needed that will become the framework used to execute development. At a high level, this plan starts out looking like this. In practice, this plan will expand and become more detailed. Post the plan somewhere that is accessible to all authors and developers that will need to contribute or view:

1. Create user stories.

2. Technical familiarization—understand the Flickr API and rules (see https://www.flickr.com/services/developer/api/).

3. Design functional requirements.

4. Build application navigation framework.

5. Build database schema.

6. Build connections to database.

7. Build connections to services (Flickr):

    a. Authentication

    b. Connection rules and limitations (3600/(total user base)/hour)

8. Implement photographer workflow:

    a. Test and troubleshoot

    b. Revise

9. Implement visitor workflow:

    a. Test and troubleshoot

    b. Revise

10. Graphics treatment (buttons, backgrounds, transitional effects).

11. Test.

12. Deploy to App Store.

13. Launch alpha users' test:

    a. 10 visitors

    b. 10 photographers

14. Revisions.

15. Final deployment and release.

Note that the item *technical familiarization* takes a very high priority. There is no better way to develop yourself into a corner than to not be familiar with the roadmap of the application you are developing against. The critical success of your app hinges on the developer's and architect's possession of knowledge of what functions and features are supported, and will continue to be supported, and the knowledge of what is coming. Now that the functional description and development plan are in place, it is time to decide on the interface model. In Xcode, when you first start a new project, you are presented with eight choices.

8.10.1 Choosing an Application to Build
8.10.1.1 Master-Detail Application
This type of template provides a framework for a master-detail paradigm of software interface. In the master-detail approach, master elements are

FIGURE 8.2   In the master-detail application, a scrollable pane on the left drives the content in the right. Tapping on an image on the left enlarges it on the right, and tapping on an image in the right-hand pane brings it to the top.

navigated through clicking and driving content through a frame positioned next to the master. This can become, then, a split screen on an iPad. Figure 8.2 demonstrates the layout of a master-detail application. The app that we are building will be a master-detail-based application.

8.10.1.2 OpenGL Game

This view allows creation of an Open Graphics Language (OpenGL) program. OpenGL is an application program interface (API), and it allows the user to create 2D and 3D graphics that are rendered at the hardware level. It was originally developed by Silicon Graphics. It is now managed by the nonprofit open technology group Khronos Consortium. Development of the standard is performed by volunteer members through a Github repository.

The OpenGL standard is comprised of an API that is very well supported and documented, and its features and functions are determined through the joint efforts of all its members, which include corporations such as Silicon Graphics®, Google®, VMWare®, Toshiba®, Texas Instruments®, and interestingly, Los Alamos National Laboratories®.

By interfacing directly to the graphics-powered unit (GPU), programmers can leverage hardware-based, graphics acceleration.

8.10.1.3 Page-Based Application

Whenever you see the o o o at the bottom of an app screen, this is a page-based application. It is very similar to a single-view application, except that it has more pages. Each pane may contain a connection to a data source and provide different information, or may provide the same information for a different configuration. For example, Yahoo's weather application allows you to create a new page for every location.

This application includes controls that allow you to create page-turning effects. However, bear in mind that just like the broken record analogy, rehashing old paradigms into graphics-based symbols is analogous to the old floppy disk icon that you see for "save file" operations in Windows. Someday, all the hard work you put into the analogy that, essentially, serves to do nothing but entertain and eat CPU cycles will be meaningless to the vast majority of people. Sliding and flipping effects are nice in that they give the user a sense of having been somewhere, and provide them visual cues and anchors into past navigation.

8.10.1.4 Single-View Application

A single-view application provides the beginning of an application that uses a single view. Comfortable in the knowledge that a single-view application can be converted to a page-based one later on, this application will begin with a single page view, and lay out a design as outlined in the functional descriptions. More on that in the next section!

8.11 CREATING THE ARCHITECTURE

This is an application of advanced-level skill needed and advanced aspects. The entire project and all the development behind it are fully documented on an online resource at http://scorellis.net/scorellient/?p=251. As the application continues to grow, it will continue to be documented. In the following sections, I walk through the basic steps needed to establish the groundwork for future application development.

8.11.1 Build Application Navigation Framework

To begin, start up Xcode. You will need a Mac of some sort to use Xcode. Later in this chapter we will explore the physical specifications needed to develop on a Mac. Sure, it may seem like there isn't much that is going into

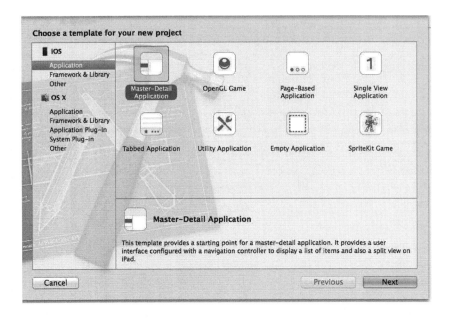

FIGURE 8.3   Choose a template for the type of program you are creating. This means making a rather important decision early on that may be difficult to change later.

your program, but you would be surprised by the tremendous amount of underlying code that there is. This is all code that you have access to and can control the behavior of, so it must be stored in textual form, and then compiled into your program when you are finished.

Selecting File from the menu bar allows you to choose New|Project. You can name your project whatever you desire; I chose "eeCardinal." When you first open Xcode, a splash page appears that contains a [Create a Project] button.

Next, we have a screen that offers us some template choices. We chose Single View application, which will later grow into a page-based app. Figure 8.3 shows the choices available in Xcode.

Click Next, and there are some self-explanatory items, such as the name of the app and your name, and file prefix. A best practice, if you like the name of your app and if you think it will be permanent, will be to use the application name for this.

* * *

In 1999 I worked on a team as an architect and was developing a new management data model. I often referred to the project as the "new data model." You cannot imagine my surprise when the developer actually took me quite literally and all of the application prefixes were NDM (New Data Model). Decisions like this are rather permanent, and going back and renaming everything can be a bit of a chore. Some decisions are best made up front. Flash-forward more than a decade: I was working with developers to produce an application that will be used to monitor the performance of a system. A similar event occurs, only now the incorrect naming scheme is strewn throughout the application in locations where industrious and more curious users can see it. Oops. And they also had one of the letters in the acronym wrong. By the time I found out, it was too late. Fast-forward 2 years and my level of involvement is such that I developed a document titled "naming conventions," which I am sharing here:

NAMING CONVENTIONS

Note: If the components that you are building are autogenerating code, please discuss any differences in formatting that it may have with the team so that any differences may be accounted for.

DOCUMENT NAMES

Documents created within the project will be given an abbreviated standard prefix eeC_.

Documents about the iPhone app piece will be prefixed with the word *eeCardinal.*

Documents about the photography on Flickr piece will be prefixed with eeC_flkr_.

PREFIX/SUFFIXING

The suffix will depend on the item being documented. For example, eeC_flickr is a document that details all the pieces in eeCardinal that provide the working functionality for inputs to the eeC data model when Flickr is called by the eeCPhotoGeoLoc function.

Avoid writing extremely long nested functions. Wherever possible, break these into multiple steps. This improves readability of the function.

Be very liberal and conversational with comments. Personality is encouraged, but avoid foul language or criticism of others.

SQL PROCEDURES/FILENAMES/FUNCTION

All SQL procedures and function names should begin with eeC_, followed by the author initials, followed by the procedure name. If an item has a multiple-word name, it follows mountain format, beginning with a capital letter. For example:

eeC_VectorGet
eeC_VectorMap

.NET Methods, Classes, Functions, etc.

File names and .NET functions, methods, and classes.

Table Names

Table names should begin with a capital letter and should not include any dashes or underscores. Column names in tables should always begin with a capital letter where possible. An exception would be where the source table does not include a capital letter at the beginning. Multipart table and field names should use the mountain format. For example:

MyTable
YourTable
HighTowerTable

Column Names

If a column contains a foreign key to another table, it should contain the table name and then the key name or whatever can best describe the field and its source table. For example:

ComplexityAnalysisViewCriteriaArtifactID

It's quite a mouthful, and abbreviations are acceptable, but they should be very clear and perhaps spelled out in a comment where necessary.

For example, the above could become ComplexityAnalysisVCArtifactID.

Key Names

This is an abbreviation of the table name that uses each letter from the first word of each part of the table name, followed by ID.

Foreign Key Names

When using a foreign key, prefix the column name with QoS_ followed by the actual key name.

Indexes

Indexes and statistics that are created for performance reasons should be created with the following conventions:

QX_keyField_

Variable Names

Variable names should begin with a lowercase letter, and should have no underscores, dashes, or special characters of any kind. Additional words should be capitalized. For example:

xRay
varscatBeginDate

QUALITY NAMING CONVENTIONS

Quality refers to the web component being built by nSerio. Quality should follow similar naming conventions where possible.

* * *

Now, you will need to configure some general options. This is metadata about the project, and also about the behavior of the project and where you will deploy it.

8.11.2 Identity

Once you have completed this screen, you are pushed to a screen that has several configuration sections for your app. Examine Figure 8.4, and note the warning.

Once you have joined the Apple® developer community, you will be able to remove this warning. Later, once this application is completed, we will pay the fee and join. Only members can deploy applications to the Apple Store. We will join so that we can walk you through the process of deployment. For now, we will just ignore this warning. The team drop-down allows you to share the project with others.

8.11.3 Deployment Info

Now, things are starting to get mildly interesting. You need to make some more choices about your application. Figure 8.5 shows the choices you have to make now.

Choose which version. For this application, we chose iOS 7 to take advantage of the newest features. Optimally, applications should have at

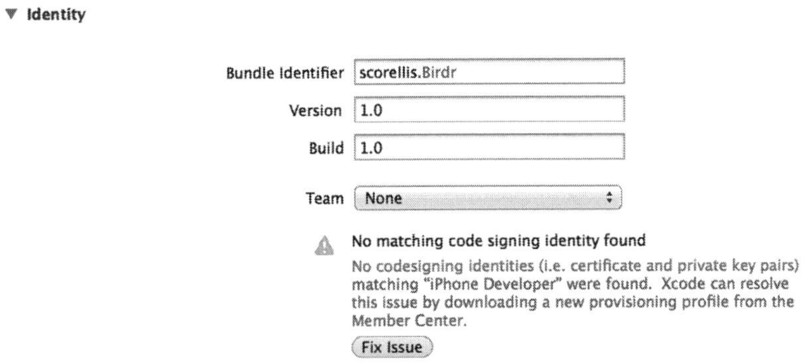

▼ **Identity**

Bundle Identifier	scorellis.Birdr
Version	1.0
Build	1.0
Team	None ⬍

⚠ No matching code signing identity found

No codesigning identities (i.e. certificate and private key pairs) matching "iPhone Developer" were found. Xcode can resolve this issue by downloading a new provisioning profile from the Member Center.

(Fix Issue)

FIGURE 8.4   Identifier information section shows a warning about the application signature.

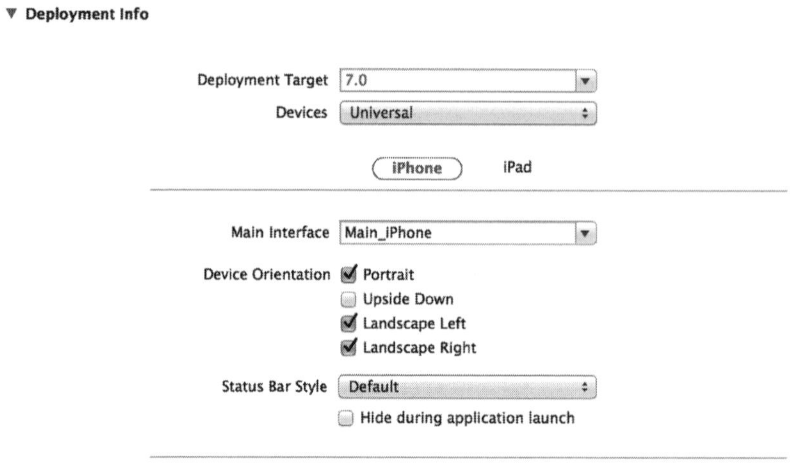

▼ **Deployment Info**

Deployment Target	7.0
Devices	Universal

            (iPhone)   iPad

Main Interface	Main_iPhone
Device Orientation	☑ Portrait
	☐ Upside Down
	☑ Landscape Left
	☑ Landscape Right
Status Bar Style	Default
	☐ Hide during application launch

FIGURE 8.5   Choose your target and orientation for deployment.

least one version of backward compatibility. For the sake of expediency and meeting publisher-driven and coauthor-mandated deadlines, this app is being built for just one version.

Caution: As always, be very careful here. If you begin to build your app on the day that iOS 8 is released, you may find that you have very few people who will be able to use your app. If you have any integration points with other services, if you are consuming anyone else's application program interface (API), you may find that the remote service is *not* compatible (yet) with the new OS.

In the devices drop-down, you can select whether the app can be both iPad and iPhone compatible. Note the [iPhone] button. This controls the orientation settings for each device. By selecting either button, you can control the orientation you will allow for each one. For example, you may have a different layout that becomes available when the iPad is oriented in landscape, but you may feel that this layout is not a good fit on a smaller iPhone screen. Below these settings, there are two more sections of interest:

App Icons

Launch Images

These will be discussed in greater detail later when designing the finishing touches on the application.

8.11.4 Framing the Navigation

We start with building the navigation for user type selection: photographer or visitor? Select buttons and save to a settings file. Start with the photographer branch of the application process flow diagram (Figure 8.1).

Main.Storyboard is a user-friendly GUI for creating things like buttons, labels, and title text. In the hierarchical tree on the left side of the application design interface in Xcode, click on it to access the design interface. Next, select the cube-looking icon from the pane on the lower right-hand side. This will give you access to items that you can drag and drop. First, build the initial welcome screen that every user will see the very first time they access the application. This interface will also be available from the settings page.

> Question: Can a photographer remain in logging mode and switch to visitor mode?
>
> Answer: Definitely! But photographers are notoriously vain and egotistical, so a photographer will only want to see her photos that she took at this location. If she has no photos, offer to log her out so that she can access the application as a visitor.

After creating two buttons and a label,

[Photographer]

or

[Visitor]

access the eeCardinalViewController.h file to declare some variables.

The standard header file looks like this, and as can be seen, the buttons I created did not autogenerate. They must be coded by hand.

```
//
// eeCardinalViewController.h
// eeCardinal
//
// Created by Scott R. Ellis on 2/8/14.
// Copyright (c) 2014 Scott R. Ellis. All rights
reserved.
//
```

```
#import <UIKit/UIKit.h>

@interface eeCardinalViewController : UIViewController

@end
```

The following is a very high level approach on application development. Actual code samples will follow in an online forum. This chapter assumes that you are an intermediate-level developer, interested in application design and the decision-making process. The building blocks of the application, while important for things like performance and usability, are subsumed by the overall design and process flow that you must architect. This is a chapter for beginning through advanced application developers, but it is not a chapter for novices. If you have never written a line of code and have never been involved in application development, you should run through some tutorials and raise yourself to the beginner level prior to building a comparable application on your own while following along with this chapter. That being said, be mindful of the following constraints when developing your application in Xcode:

- The amount of memory you have available is different in iOS 6 and 7, and there is no page file.

- Design the app to be used on mobile, but be mindful of letting the user know his usage.

- "Building" creates a lot of extraneous files that of course take up space. To get rid of them, use the "clean" feature.

- Enable "backgrounding." This application will need to be able to run in the background when in photographer mode.

- Coding for device orientation is not as simple as just checking off the tick box. You'll need to do some work to get it to work! You'll need to implement some code for additional device orientations.

- Create the following icons:

  - iPhone: Retina display (2×), 120 × 120 pixels.

  - iPad: Non-Retina display (1×), 76 × 76 pixels.

  - iPad: Retina display (2×), 152 × 152 pixels.

- Create the following launch pages:
  - iPhone: 3.5-inch Retina display (2×), 640 × 960 pixels (portrait).
  - iPhone: 4-inch Retina display (labeled as R4 in Xcode), 640 × 1136 pixels (portrait).
  - iPad: Non-Retina display (1×), 768 × 1024 pixels (portrait).
  - iPad: Retina display (2×), 1536 × 2048 pixels (portrait).

- Set up and explain unit testing procedures.

- That little triangle next to the folder name is called a disclosure triangle. When you click it, the contents of the folder are revealed.

- View build logs by selecting the Log navigator; use Navigator selector or choose View ⇒ Navigators ⇒ Show Log. On the Navigator Selector, it looks like a little cartoon speech bubble.

- Frameworks in Xcode are very important. Particularly central to the Apple iPhone/iPad SDK, the UIKit framework provides just about everything needed to build the graphical user interface. I will provide some level of detail to help familiarize you with design patterns and the underlying paradigm that serves as a functional foundation for developing an application. This will be pretty high level, with an occasional high dive down into the deep end to explain things that you may find particularly tricky. The point of this chapter is not to teach you everything you need to know to create an app, step-by-step. Rather, it is to provide you with the stepping-stones needed to guide you through rough, churning waters. I can't promise you that the stones won't appear to be partially submerged, or covered with slippery river algae, but I can promise that I will make the source code for this project available at http://www.scorellis.net/steppingstones.

  - NIB (with its corresponding file extension .xib) is an acronym for next interface builder.

8.11.5 Workflow

Just like building a building, applications have foundations, frameworks, and toolkits. Xcode is no different, and as we dive into development, we must first understand how all the pieces will fit together. We have, at this point, a list of pages on our storyboard, and we can attach frameworks to each. In Xcode, a framework provides a grouping of generic functionality. Much of the core features are those of similarity to your iOS-powered devices.

Coding and Application Design Tip

One quick note about pop-ups—in my ever so humble opinion, pop-ups are a crutch and should be avoided at all costs. You may find yourself creating an application that allows a user to delete all of his contacts, and then you ask, in an annoying pop-up: "Do you really want to do this?" So, there actually is such a thing as a dumb question, and asking a user if he really wants to do something rises to the pinnacle of developer arrogance.

"But wait!" you say. "If I can't seek confirmation of something, then what do I do when someone unintentionally deletes his or her information?"

There are multiple problems with this: (1) you have to code a model for it and (2) if the user is really in a situation where he has unwittingly selected to delete or destroy information and was unaware of it, then you need to take a long, hard, deep look into your soul and figure out how you allowed that to happen. For example, if shaking the phone is a function to have people delete things, and you pop a modal for confirmation, then you need to find a different way of letting people delete, because every time they accidentally bump their phone, they are at risk of deleting data.

The evolution of the *confirmation window/modal/annoying pop-up* is clear. In real-world mechanical and electrical solutions, humans are moving through three-dimensional space and they don't always have perfect awareness of their position. Sometimes, they (we) bump into things quite accidentally. So, switches and levers will have what is called a lockout mechanism of some kind. A lockout or safety is a thing that you have to do in order to mechanically enable the switch.

Given all the nifty controls in Xcode, you could certainly do something similar to that, and have a slider switch that must be moved that then enables a "Delete" button. I can think of an even better solution, though. Can you?

The magic of computer systems is that you can store data. You can flag things. For example, when the user deletes all of his contacts, you simply set a bit column in the data table as isDeleted = TRUE and then allow the user to undo an operation. Then, on a settings page, you can allow someone options for data retention, such as "clean up deleted data on shutdown" or "clean up deleted data daily" or "retain undo data" or "retain undo data for 00 minutes," and allow the user to set the time.

8.11.6 Object Model

Decide on an object model, using the real-world interstices with the application as our starting points for the objects.

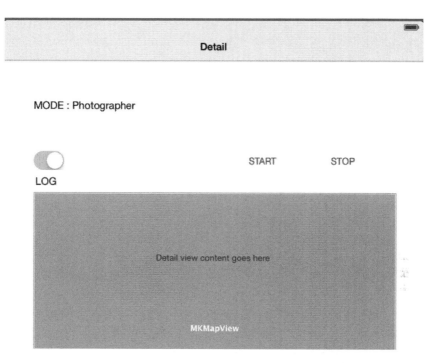

MODE : Photographer

START   STOP

LOG

Detail view content goes here

MKMapView

FIGURE 8.6 Mock-up from the eeCardinal application. In this workflow, in order for the photographer to begin logging, he or she must slide the slider to the left to enable logging, and then click Start.

1. Viewing:

   a. Getting location.

   b. Getting photos.

   c. roleModal—what role is this user? Different users have different modalities.

2. Moderating

3. Logging—Figure 8.6 pictures how the user must log in in order to enable this feature.

4. Synchronizing

Each of these four areas represents natural delineation within the development framework, with the capacity for each piece to be executed and then brought together in a natural framework. Each can then be thought of as having separate paths to completion. Many applications have multiple facets, and represent multiple stories that all come together to form a single epoch.

At this stage, you have perhaps begun documenting and planning out an app of your own. Now it's time to get something built that is not sophisticated, is completely unimpressive, but *works*. It is very skinny, but it represents the basic idea of what you are trying to accomplish. If you work for a software firm, you might want to research and understand the hackathon.

8.11.7 Hackathon

Nobody is certain where the word came from, but it is a portmanteau of the words *hack* and *marathon*. Traditionally, the word *hack* never meant "crack," which is just one meaning of the word. In the hackathon sense, the word simply means to hack something together. Several of the chapters in this book (including this one) are the result of hackathon-like activity on the part of the authors. When you see sidebars, this is almost assuredly a "hack" from some other article that one author or another had that was offered up and inserted to improve content.

How do you conduct a hackathon? It's easy. The rules are simple. Giving yourself no more than 48 hours, create a working model. What is the least you can do in that amount of time?

For this sample, we gave ourselves far less than 48 hours, and it did not result in a working model, but did come up with some nice screenshots and a flowchart of how the app will work. To be fair, we were writing a book at the same time.

8.11.8 Initial Model

See Figure 8.7.

8.11.9 Interface Preview

Once a user has installed eeCardinal, and it identifies her location, geo-tagged photos from Flickr are presented. Pictured in Figure 8.8, the panel on the left allows scrolling, and when tapping on a thumbnail, the full photo appears. By virtue of being selected, it vanishes from the thumbnail and a new one takes its place.

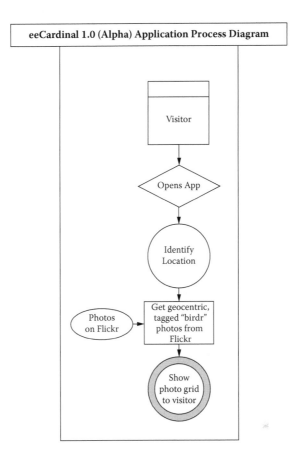

FIGURE 8.7   The right half of the project lends itself well to rapid development.

FIGURE 8.8   In this mock-up of eeCardinal 1.0, photos were taken on the Chicago River near Jackson Street displaying to visitors in the area.

8.11.10 Conclusion

The remainder of this project is available at http://scorellis.net/ scorellient/?p=251. Here, there will be links to the code repository, step-by-step directions, and you may join in the conversation. Whether or not this will become an open-source project is entirely dependent on the number of photographers that get involved and the interest of the coding community.

III

Mobile Services

CHAPTER 9

Mobile Standards

Scott R. Ellis

CONTENTS

9.1 HISTORY OF MOBILE

The history of mobile cellular-based communications began in the early 1900s, but the history of wireless, long-distance communication has roots that can be traced back much further. To understand existing standards that are currently in place in wireless communications, it is useful to examine and understand the history of mobile and all of its generations.

From its earliest days, the mobile industry faced challenges related to regulatory compliance and standardization. The industry lacked standards and standardization. Companies that built the mobile technology wished to corner the market. They wanted to gain an appreciable market share and dominate their competition. This history of the delays in coming to market, and the slowness of adaptation, and how standards came to bear, over time, deserves an in-depth examination.

This overview of how mobile technology and telephony developed examines the history as it pertains to the standards and technology of today, and follows as direct a path as possible through the development of the technology. From the inventors and enterprises that worked to stand up the industry, to the organizations that worked to regulate and standardize it, this chapter attempts the most thorough treatment of this subject matter as is possible in the pages permitted.

Certainly, the earliest suggestions of remote communications probably have to date back to prehistoric visions and peyote-induced (far-seeing) hallucinations. One wonders what the ancients would think now, when we hold a slim piece of plastic, metal, and glass and receive distant images and audio. According to Ray Kurzweil, Google® director of engineering, by 2029 computers will match human intelligence. Nanotechnology will create virtual realities within our nervous systems.* You can bet it will be wireless enabled, and tapped in to the Internet, networked, and chockablock with advertising. What, one wonders, would a collective mind entertain itself to accomplish? Will the concept of the individual perish? Would we "unplug" from one net and plug in to another for recreation, living double lives? Most importantly, how will we write apps for it, and how will it be monetized? And if so, will it be monetized? This brief history of wireless will endeavor to provide a foundation for understanding these questions, and it will begin in the year 1792, with the first overland, semaphore-based "wireless" communications.

9.1.1 1792: First Successful Wireless Optical Telegraph

Certainly long-distance signaling existed prior to this, but credit for the ability to communicate entire messages in complete sentences over long distances in a useful way goes to French physicist† Claude Chappe and his brothers. They succeeded in covering France with 556 stations, networked together in a fully operational, mechanically optical-based semaphore line. With a reach of 4800 kilometers into neighboring countries of Holland, Italy, Germany, and Spain, it was used by Napoleon to coordinate his empire. It remained in service until 1854. This mechanical system allowed an operator to stay inside of a hut and work a mechanical system to position large rods on a tower above the hut. Neighboring huts would monitor each other through telescopes, and would then relay the message to the next hut. Similar systems were built in Britain, spurned by the fractious nature of Anglo-Franc relations at the time and a warmonger named Napoleon.

Chappe had invented what would be called a semaphore telegraph. Though he eschewed the name *telegraph*, this is very likely the first use of

---

* As a Google director of engineering, Ray Kurzweil works to improve computer understanding of natural language. He is the author of *The Singularity Is Near: When Humans Transcend Biology* (New York: Viking Press, 2005), and he is working to reverse engineer the human brain.

† Foy-les-Lyon, Sainte, Marcy Sur Anse, Jonquieres, Haut-Barr, Annoux, and Pleumeur-Bodon, "Les Télégraphes Chappe, l'Ecole Centrale de Lyon," [French article], *Le Télégraphe Chappe*, Cedrick Chatenet.

the word. The Chappe brothers had called it a tachygraphe. The root *tachy* means "rapid," or "accelerated," and *graphe* means "to write," or "that which is written." The system wasn't literally "that which writes fast," because it was actually a very slow process of writing. Rather, the Chappe brothers probably were thinking of it as "accelerated communications." Even though the process of interpreting the messages was slow, the communication itself relayed and traveled much faster than the fastest communications of the day, which would have been horseback. Operators need not even know the content of the message; they simply needed to mimic the semaphore positions in order to relay them.

During this time it came to pass that the Chappe brothers were introduced to French diplomat Comte André-François Miot de Mélito (1762–1841). In his role as Comptroller General of the Administration of Military Affairs, he suggested to them that they change the name to *telegraph*, "far-writing." It is unclear if Comte invented this term, or if it was already in use and he applied it to the Chappe invention. In his memoirs, which can be found online, he describes how he suggested this to the Chappe brothers when he was introduced to them by the "famous artist David," most likely a reference to Jacques-Louis David, a prominent French painter at the time.* Judging from the pride by which he describes how it later became "so to speak, a household term," it is a fair assumption that perhaps he did coin the term.

The telegraph system invented by the brothers Chappe (as shown in Figure 9.1) was also known as the Napoleonic semaphore. It used a system of lever-operated blades, which would be positioned in a prescribed arrangement to signal individual letters of the alphabet. It remained in use until the 1850s, when it was replaced by the electric telegraph. It was far more efficient than postal riders, but in terms of the security of the message, the probability that ciphered communications would have been used is high, but this author could find no reference to them.

Other inventions that followed a similar concept included the Murray shutter system, which was deployed in the United Kingdom from 1795 to 1816. Note that the shutter system is an excellent example of a multi-bit communication system, with shutters configured in a 6-bit arrangement (as shown in Figure 9.2). Later, the telegraph system would come to adopt a single-wire system using Morse code, which was far more cost-effective than other multiwire systems at the time. It would be many years

---

* Memoirs of Count Miot de Melito, Minister, Ambassador, Councillor of State, 1881, p. 44.

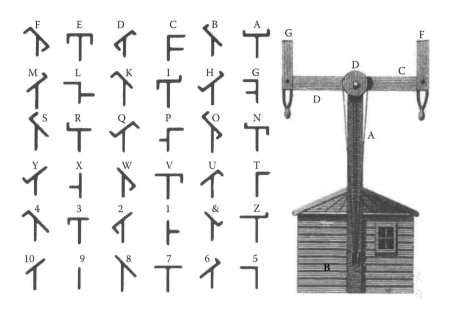

FIGURE 9.1 Original patent submission drawing. (From Rees's *Cyclopædia*, Plates Vol. IV, "TELEGRAPH," Figure 4.)

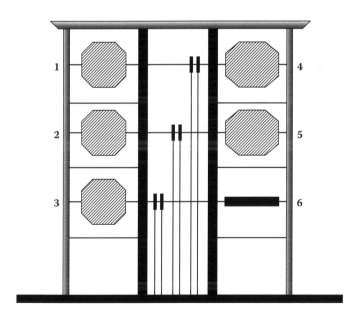

FIGURE 9.2 The Murray shutter system, deployed in the UK as a contemporary system to the Chappe system, introduced a method of communicating that bore a strong resemblance to a multibit form of communication. It was later replaced with a three-armed system similar to Chappe's levers.

before an actual multibit system would come into use in the teletype, which automated information delivery for the first time.

9.1.2 1837: Electrical Telegraph

In 1825, upon the death of his wife while he was out of town, Samuel Morse, a painter, vowed to develop a form of long-distance communication. While there were numerous other telegraph inventions that were patented at the same time, Morse's invention was the simplest and easiest to use, and by 1861 it had connected the East and West Coasts of the United States.

9.1.3 1876: Alexander Graham Bell Patents the Telephone; 1880: Invention of the Photophone

Alexander Graham Bell and Charles Sumner Tainter carry out the first wireless phone conversation. Slightly ahead of its time, the invention was not seen as useful at the time because it required sunlight to operate—the light bulb was patented on January 27 of that year.[*] Bell and Tainter were on the right track, but this is a good example of an invention being ahead of its time.

9.1.4 1908: Wireless Voice Communication

Professor Albert Jahnke of the Oakland Transcontinental Aerial Telephone and Power Company schemed to sell stock in a company that claimed to have developed a wireless telephone that, according to one magazine, was "not wireless and would not telephone." The company was put on trial for fraud, and while the charges were dismissed,[†] the company ceased to do business in July 1908. In one accounting, in print, a reporter for the *Pacific Telephone Magazine* clearly states that the invention was fraudulent, and consisted of a tool chest-sized box placed against a rather thin wall and another box against it in the other room. Picking up the box and turning it in the opposing direction broke the induction connection and the device would not work. According to the reporter, this did not seem to matter to many potential investors.[‡]

A separate account by Frank Comody, who testified under oath, also from July 1908, claimed to have seen a demonstration of the device in

---

[*] U.S. Patent 223,898, issued January 27, 1880, to Thomas Edison for an electric lamp and manufacturing process.

[†] *San Francisco Call*, vol. 104, no. 37, July 7, 1908

[‡] *Pacific Telephone Magazine*, vol. 2, no. 1, July 1908.

1904 where clear speech was communicated over a distance of 7 miles, that it did so in a wireless manner, and he suggests that he believed the underlying technology was sound, and that if the device were powerful enough, it could communicate over a distance of 1000 miles. Comody had undoubtedly been invited in as an "expert" witness, but it is not clear from court papers for which side. It is doubtful that the prosecution would have presented testimony damaging to their case. It is equally doubtful that an electrician, one who mostly runs wires and installs items that use electricity, would have the advanced knowledge necessary to testify to advanced matters regarding electromagnetism and the complex communications circuitry necessary for a wireless voice exchange.

Regardless of whether the device worked or not, or if Comody, an electrician, had been the victim of an elaborate fraud, Jahnke did not proceed with production; the company dissolved.* At best, it may have worked. At worst, if it did, it is ranked as an epic fail from a business perspective.

9.1.5 1918: German Wireless Telephony

As early as 1886–1887, Heinrich Rudolph Hertz, a German physicist, had built a wireless transmitter and receiver. Hertz considered his invention to be one of pure science, however. He merely sought to clarify and expand upon the work of James Clerk Maxwell, who had developed an extraordinary set of equations that pertained to electromagnetism and light. Hertz actually saw no practical application for his invention, which is demonstrated in Figure 9.3 to be that of an electromagnetic coil[†]—one of the, if not the most important, components in modern communication broadcast systems. When asked about his work, he would say, "It's of no use whatsoever…. This is just an experiment that proves Maestro Maxwell was right—we just have these mysterious electromagnetic waves that we cannot see with the naked eye. But they are there."

9.1.6 1894: Marconi Experiments

Following Hertz, Guglielmo Marconi (1874–1937), a physics student who had seen Hertz's electromagnetic waves experiment reproduced, immediately saw a technology that would lead to his first patents. Marconi is often credited with at least co-inventing the radio as a form of communication,

---

* Electrical Journal, vol. 61, p. 589.
† Heinrich Rudolph Hertz, *Electric Waves: Being Researches on the Propagation of Electric Action with Finite Velocity through Space*, p. 34.

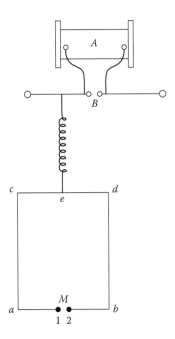

FIGURE 9.3   An excerpt from Hertz's experiment, which proved the existence of electromagnetic waves, and the ability to accurately detect them.

and he shared a Nobel Prize for his work in 1909. Marconi began with two large poles. By connecting one of the poles to the ground and the other to a suspended wire, he initially was able to reach a little over a mile in transmission.[*]

By 1896, the system that Marconi was working with included well-known components. His transmitter included the Righi oscillator, but Marconi had figured out that a larger spark gap would allow him to increase the range. The coherer served as a receiver. The coherer had been invented by Edouard Branly in the 1880–1890 time frame, and worked on the principle that when an electromagnetic field is applied to a vacuum tube filled with metal filings, the resistance of an electrical current being passed through them would decrease. The coherer used by Marconi would have looked somewhat similar to the device shown in Figure 9.4.

For an antenna, Marconi used a vertical antenna attached to ground. This style of antenna, invented by physicist Alexander Popov, used a coherer connected in a circuit to a battery and a pair of electromagnets. The coherer would close the connection when radio waves were detected,

[*] Michael Friedewald, *Journal of Radio Studies*, vol. 7, no. 2, 2000, p. 2.

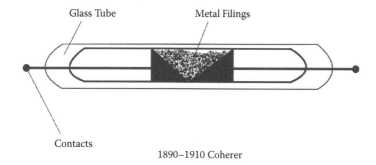

Glass Tube   Metal Filings

Contacts

1890–1910 Coherer

FIGURE 9.4   Early coherers contained a chamber filled with metal filings, which would "cohere" and increase in conductivity in the presence of an electromagnetic wave.

effectively communicating a signal. A Morse writer used to record the message completed Marconi's first system. Marconi made additional improvements, and ultimately completed the first successful transatlantic communication after considerable long-distance experimentation from 1901 to 1903. Interestingly, part of the reason for Marconi's success may be attributed to the fact that while electrical communications were under government control, Marconi had incorporated and retained control of all of his transmission stations.[*]

9.1.7 German Wireless Research

Early on, the politics of wireless communications was beset with political motivations—England and Marconi wished to retain a massive monopoly of control over the industry, and refused to join in any international agreement regarding interoperation of wireless operators and stations. Marconi dominated the maritime industry, and his operators were instructed that under no circumstances should they speak to any competing operators. As early as 1897, possibly in response to the successes of Marconi, Germany began to ramp up research and development in the technology. Voice transmission experimentation had been taking place, but it was still considered a novelty. Industry ruled the day, and military, and their highest priority was the ability to modulate signals into narrow bands to prevent easy interception. By 1912, the critical importance of wireless telegraphy to maritime industry had become apparent. Ships were able to call ahead and inform the port of arrival, and most importantly, as in the

---

[*] Michael Friedewald, "The Beginnings of Radio Communication in Germany, 1897–1918," *Journal of Radio Studies*, vol. 7, no. 2, 2000, p. 2.

case of the *Titanic*, send distress signals. Increases in shipping in the early 1900s increased the requirement for greater security of communications. With thousands of ships on the lines, overlapping communications posed a very real problem. Beginning in 1918, Germans would be first to test, on their railroad system, wireless telephony on military trains between Berlin and Zossen.

9.1.8 1924: Public German Wireless Telephony

Public trials started with telephone connection on trains between Berlin and Hamburg.

9.1.9 1926: Wireless Mobile in the Media

German artist Karl Arnold drew a cartoon demonstrating mobile phones. The cartoon, captioned "wireless telephony," was published in a satirical magazine, *Simplicissimus*. The cartoon shows what appears to be commuters with wireless phones that look not unlike a miniature handle-set of a turn-of-the-century (1900) phone, with a wire that connects the handle-set to a small box in the breast pocket of its users.

9.1.10 1926: Telephone Service in Trains

Telephone service of the Deutsche Reichsbahn and the German mail service on the route between Hamburg and Berlin were offered to first-class travelers.

9.1.11 1940s

Mobile telephones for automobiles became available from some telephone companies in the 1940s.

9.1.12 1947: FCC Involvement

The Federal Communications Commission (FCC) limits the amount of frequencies available in a particular area. These limits made only 23 phone conversations possible simultaneously in the same service area. What was the range of these frequencies? Had digital technology and switching even been considered? How was the signaling equipment being constructed? What would have been the technology capability at the time to handle thousands of users? What would the military and security consequences have been? What was the capability of enemy signal intercept? Some people may want to disparage the FCC, but what was the rationale for the decision to be so limiting?

9.1.13 1947: Police Wireless

Bell Labs introduced cellular-like technology for police communications.

9.1.14 1956: Sweden

The Metropolitan Transit Authority (MTA) system was the first system with an automated switching network for vehicles. The system's technology consisted of rotary dial equipment. The phones could also be paged. To receive incoming calls, an operator would first have to figure out where the car was located. Little information is available about the MTA generation, and it only had a handful of subscribers.

9.1.15 1968: FCC Changes Its Position

In 1968 the FCC reconsidered its position, stating, "If the technology to build a better mobile service works, we will increase the frequencies allocation, freeing the airwaves for more mobile phones."[*] Clearly it had not been impressed with previous attempts. It is the duty of organizations such as the FCC to ensure that the airwaves are not reduced to a jumbled, incomprehensible mass, which is surely what would ensue in the absence of regulation. The theory of the tragedy of the commons is all too real, though perhaps it should more aptly be called standard human behavior in a competitive environment.

9.1.16 1973, April

Dr. Martin Cooper allegedly places a call to his adversary/nemesis at Bell Labs, one Joel Engel.

9.1.17 1975 Patent

Dr. Martin Cooper for Motorola

US03906166

September 16, 1975

Radio telephone system

Inventors: Martin Cooper, Richard W. Dronsuth, Albert J. Mikulski, Charles N. Lynk Jr., James J. Mikulski, John F. Mitchell, Roy A. Richardson, John H. Sangster

---

[*] Federal Communications Commission (FCC) Ruling 13, FCC 2d 420.

9.1.18 1977

Both AT&T and Bell Labs entered into a prototype phase of cellular communication systems. Within a year, both companies had begun public trials of the new system in Chicago and Newark, New Jersey.

9.1.19 1979

The first commercial cell phone system commences operation in Tokyo.

9.1.20 1981

The cell phone is not a technological destination that we have arrived at—it is merely the beginning, and readers are encouraged to stretch their minds, and not view the cell phone as a platform for making a quick buck building the next great app, but rather as the beginning of human convergence. Technology and humanity are racing at breakneck speeds toward a technical and human convergence. Already, far more of my communications are conducted electronically. As the harness of technology broadens its scope, and deepens its reach, where will that leave human interaction? And how do we leverage the knowledge of this direction when designing and building applications? Does the application or technology you want to build support this direction? Our devices exist to help us make better decisions, and to make more of them.

Bell and Tainter invented and patented the photophone, a device that used wireless beams of modulated (pulsing light) to carry conversations wirelessly. In the 1880s, there existed no utilities to conveniently plug in to for electricity. The laser, so essential to fiber optic communications, would not be invented for another 80 years. In fact, 40 years would pass from the time the first scientist theorized that lasers even existed. In 1880, however, there was no practical way to implement the device. What would one do? A portable generator to generate electricity was about as large as a house, and the device itself was far from pocket sized. The photophone also required a clear line of sight, similar to devices used today: infrared networking devices and fiber channel.

Similar to free-space optical communication, the photophone also required a clear line of sight between its transmitter and its receiver. It would be several decades before the photophone's principles found their first practical applications in military communications and later in fiber optic communications.

The history of cellular telephony intertwined inseparably with that of wireless communication. Consequently, the fact that Alexander Bell achieved first wireless communication that used technology to convert a distant semaphore into words should not surprise anyone. Prior to this, the military had used a wide variety of signaling methods, from balloons raised to cannon shots and flares.

9.2 STANDARDS

From the earliest days, standardization was necessary. In the late 18th- and early 19th-century France, the semaphore telegraph lines all used the same form of communications to communicate. The Chappe system was the *standard* model, and it embedded both an equipment and a communication encoding standard, as developed by Chappe. Other systems employed standards as well. In fact, each semaphore line in each country employed separate semaphore systems, with the exception of the French lines, which expanded into other countries and carried forward the same standard of communications as the originating lines. In fact, another word for *standard* is *generation*, and it can be thought to be an expression of interoperability between current distributions of an underlying mechanized communications substrate.

The previous section of this chapter concerned itself with the development, over history, of that mechanical/electrical substrate in order to demonstrate the age-old adage that you "can't know where you are going if you don't know where you have been." As the story progressed into the modern era, it interleaved more details of each relevant standard. Finally, it left off with development of the 3G standard, mentioning it as basically a footnote.

Now more than ever, we still have no less of a requirement for standardization. Cell phone manufacturers must follow radio signaling guidelines. This is not a complete list, but rather is representative of past standards and only comprises a very small slice of the overall alphabet soup of this industry and its plethora standards. As cell phone use increases and becomes more prevalent, and as it becomes ever more entrenched in our culture and increases in business relevance, it also increases the tendency of the quality of service capabilities of the underlying technology to diminish, inciting a need for new technology that improves the speed data buffering, increases bandwidth data capabilities to reduce network congestion, interleaves intervals, and waits for resource allocation. All of these factors, added together, and incurred with processing overhead and network latency, create an overall perception of latency that the user perceives as a poor experience.

Some online sources suggest that the quality of voice conversations diminishes as the distance from a cell tower increases because "as they move farther from the base station, the signal will take longer to arrive." This is perhaps a misunderstood oversimplification of the very real results of signal dissipation due to the effect of an ever-expanding spherical radiation (inverse square law), which may increase latency and packet loss due to retransmissions. Radio transmissions are a form of electromagnetic radiation.

Electromagnetic waves travel at the speed of light. Certainly, the signal itself takes no longer to arrive; it takes longer to complete a successful transmission. Light (in a vacuum) travels at 299,792,458 meters per second. So in a millisecond, which is 1/1000 of a second, it will travel 299,792.458 meters—186.2824 miles, or 186 miles 495 yards 5 inches. The range of most cell towers is probably just a fraction of that. The complete range of most towers (about 6 miles) can be traversed by a radio wave in about 3/100 of a millisecond.

Latency, typically, is measured in whole milliseconds. In computer systems, such as database and network systems, less than 1-millisecond latency is absolutely fantastic and is just as good as zero latency.

Tip: To get a good understanding of latency, try photographing anything that is in rapid motion, such as children, or perhaps a friend's fleeting smile. Most off-the-shelf consumer-grade digital cameras have latency upwards of 1800 ms, which is almost 2 seconds. The very best digital cameras have a latency of between 30 and 50 milliseconds, which is usually fast enough and unnoticeable. This latency is called shutter lag.

Latency can be thought of as the total amount of time from the time that a request for information completes to the time that it is received and in use. For example, if you say hello to someone on the phone, he or she isn't actually hearing the word as you say it. There is a delay, and it can be on the order of hundreds of milliseconds and, in a worst case scenario, seconds. Latency is a measurement, and with respect to mobile transmissions, the mobile communication standards are mostly concerned with the length of time from when one frame of information is sent from your cell phone to when it is received and heard by the recipient. Things like internetwork latency must also be considered, and this accounts for the delay between callers when they are separated by many disparate systems. Cellular network operators, then, have the difficult goal of balancing quality with latency. Increasing the quality of transmissions increases the latency, because fewer packets are dropped. Decreasing the quality decreases latency, because the system is less concerned about whether or not the entire packet stream was properly transmitted ("Can you hear me now?").

9.2.1 The 3rd Generation Partnership Project (3GPP)

Not to be confused with 3GPP2, 3GPP is a collaboration that began in 1998 between two groups of telecommunications associations, Nortel and AT&T, aka the Organizational Partners. Within a couple of years the partnership had expanded to include additional telecommunications members, and its scope became worldwide. 3GPP2 is a different partnership project for CDMA2000, in which a standard is built upon the earlier 2G CDMA work. The following systems are extant under the 3GPP governance. 3GPP, as an organization for developing standards, was certified by the American National Standards Institute.

Generally speaking, all cell phone technologies can be lumped under one standard or the other. 3GPP is the governing body of all technologies implemented on the Universal Mobile Technology System (UMTS) or GSM system. 3GPP2, a sister project to 3GPP,[*] provides governance on all things CDMA related. GSM is the most widely implemented system, used exclusively in most of Europe. Many countries, such as China and America, use both. The 3GPP2 equivalent of 3GPP is the 3G network, and is in use by American companies such as Qualcomm, U.S. Cellular, Lucent, and a few others. From publicly listed meeting and conference attendee lists, as listed in publications, support and participation of major industry players in the 3GPP2 appears to have radically reduced in the past decade.[†]

3GPP2 participating organizations in 2002 were as follows:

1. Cisco Systems

2. CommWorks

3. Ericsson

---

[*] See http://www.3gpp2.org/public_html/misc/abouthome.cfm.
[†] See http://www.3gpp2.org/Public_Html/Summaries/.

4. France Telecom

5. Fujitsu

6. Hitachi

7. Hyundai SysCom

8. KDDI

9. LG Electronics

10. Lucent Technologies

11. Motorola

12. NEC

13. Nokia

14. Nortel Networks

15. Qualcomm

16. Samsung

17. SK Telecom

18. Sprint PCS

19. Winphoria Networks

3GPP2 participating organizations in 2013 were as follows:

1. Alcatel-Lucent

2. Kyocera

3. Qualcomm

4. Sprint Nextel

5. U.S. Cellular

6. ZTE

Alternatively, just one member organization of 3GPP lists hundreds of participating companies, and represents the complete who is who of telecomm.

9.2.2 Global System for Mobile Communications (GSM)

GSM was created by the European Telecommunications Standards Institute (ETSI). It was widely used. Its creation was the by-product of a work that began in 1981 by the European Conference of Postal and Telecommunications Administrations (CEPT) in an effort to standardize and deploy a standardized, cellular-based telephony system across Europe. This work began in 1981, and by February 1987, work on the standard had been completely ratified. The most important thing to note about GSM is that it leverages a method called time-division multiple access (TDMA). TDMA is a method to access a channel for networks that reside on a shared frequency band. It works by dividing signals into various time slots (as illustrated in Figure 9.5). In other words, many users transmit in rapid succession of one another, and this allows many users to share the same frequency band without saturating the entire allotted channel.

- General Packet Radio Service (GPRS). This packet-oriented data service was introduced by ETSI and is used in 2G and 3G systems. This standard is based on a best effort approach, which means that no quality of service guarantee or capability exists. Basically, it means there is no guarantee of bit rate or delivery time. The experience of the user will be variable, depending on network load. It extends GSM capabilities and allows for services such as P2P Internet networking, push-to-talk, and short and multimedia messaging services (SMS and MMS).

- Circuit-switched data (CSD). CSD is the first technology used behind time-based switching in the GSM network, which was TDMA based.

  Before development of CSD technology, 2.4 kbit per second modems were used to transmit data, by a modem either built in to the phone or attached to it. CSD provided the underlying technology used by TDMA and GSM systems. Ultimately, with the introduction of digital transmissions, cell phones were able to nearly directly interface with the underlying equipment.

- High-speed circuit-switched data (HSCSD). HSCSD provided each device with multiple time slots, which improved performance and provided faster data rates. The additional channels also provided better tower handoffs and transitions when in the presence of multiple towers, and enabled smoother load balancing and a better user experience when traveling.

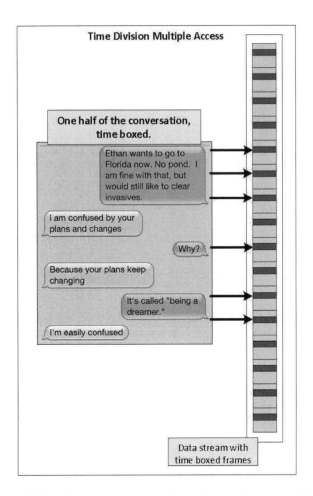

FIGURE 9.5   TDMA slices data into preallocated time frames, and intersperses the communication with other broadcasts. Data compression allows for the seemingly seamless transmission of voice conversations, as bit streams compress into and expand from the allocated time slots.

9.2.3 Enhanced Data Rates for GSM Evolution (EDGE) (E-GPRS)

EDGE introduced higher data rates and maintained backwards compatibility to core GSM equipment. It also provided backwards compatibility to GSM.

- EDGE (EGPRS). Requiring no actual hardware changes, EDGE equipment could coexist peacefully within the distributed shared memory (DSM) network. By installing EDGE-compatible transceivers, the networks can easily be upgraded.

- Evolved EDGE (EGPRS2B). Evolved EDGE introduces additional techniques to reduce latency. It is strictly a software update. Evolved EDGE also introduced the capability for a single mobile phone to transmit and receive on two simultaneous frequencies, providing that the phone has been equipped with two antennas.

9.2.4 Universal Mobile Telecommunications System (UMTS) (Revolutionary 3G)

UMTS, a third-generation (3G) system based on the 3GPP GSM system, introduced the concept of authentication of users via a subscriber identity module (SIM) card. It specified a completely new network. It introduced frequency-division duplexing (FDD), a technology whereby the transmitter and receiver operate on different frequencies. Its standard for communication, wideband code-division multiple access (WCDMA), replaces TDMA as the underpinning technology of communications management in air-based transmissions.

- High-Speed Packet Access (HSPA). HSPA combines two mobile telephony protocols: High-Speed Uplink Packet Access (HSUPA) and High-Speed Downlink Packet Access (HSDPA). This blending of technology is applicable only to networks that use the WCDMA (UMTS) protocols. It improves the performance of 3G networks. Evolved HSPA (also known as HSPA+), released late in 2008, allow speeds of 168 Mbit per second down and 22 Mbit per second up.

- Long Term Evolution (Release 8) (LTE). LTE is the continuation of the upgrade path specified by 3GPP. It allows for greater throughput than HSPA+. Work on the standard began with 3GPP in 2004 and resulted in a completed release (8) of a specification in March 2009.

- LTE-Advanced. LTE-Advanced is the next generation in the GSM upgrade path, and it follows EDGE, UMTS, HSPA, and HSPA+. LTE will bootstrap up on the heels of HSPA and will likely be the predominant technology of the next decade. At press time, the LTE standard still falls short of the complete 4G specification as published, but in terms of marketing forces, it is "close enough." Due to the advancements in 3G technology brought to the market by HSPA+ and LTE, at least one international body of governance, the International Telecommunications Union (ITU), stated in a press release dated December 6, 2010, that it had evaluated LTE-Advanced and that the technology provided such an advanced improvement to global wireless broadband communications that it should be considered 4G. In the same breath, it acknowledged that the term was undefined, but also recognized the existing 4G technologies as being of significant incremental breadth of advancement to the existing 3G systems.

9.2.5 5G—Beyond 2020

This chapter closes with a brief discussion on the future of mobile. 5G is not based on any officially recognized publication and, at this time, is comprised of mostly speculation based on industry trends and predictions for the future of communications technology. If the history detailed in this chapter is any indication of a pattern, it demonstrates that each technology undergoes major upheaval every 50 to 60 years, with major iterations within each branch occurring every 10 to 15 years. This is a pattern that extends all the way back to Chappe's system of optical semaphore, and in the spirit of optics, it is also clear there is a pattern, which suggests that the technology that will be in use, and widespread, reveals itself as early as 120 years before it gains complete saturation and acceptance. The first wireless voice conversation occurred in 1880 with the photophone, and by the turn of this century cell phones were on secure footing in culture.

There is room for speculation here. Already, smartphones have little need for cell towers to enable the capability for live video conferencing. Improvements in WiMAX delivery and in Li-Fi, which is the use of light-emitting diodes in networking communications, may satisfy the ability to deliver publicly available broadband communications for a lot price, possibly with screw-in light bulbs serving as the repeater for these systems. In a day and age when a screw-in LED light bulb, which can be purchased at your local hardware store for about $20, can be made to change colors by remote control, it's not inconceivable that such a device would be networkable. If history is any indication, one possible promise of 5G is this: the mobile infrastructure is aging and cannot keep up with the demands of users. A disruptive technology is sure to emerge that will see the abandonment of the tower-based technology of the 1980s, and that this will happen by 2030.

9.3 DEVELOPMENT OF GOVERNMENT STANDARDS

Government regulation and control of the airwaves has its roots in the earliest mobile efforts. As early as 1947, the FCC began to limit the amount of frequencies available in a particular area. These limits made only 23 phone conversations possible simultaneously in the same service area.

What was the range of these frequencies? Had digital technology and switching even been considered? How was the signaling equipment being constructed? What would have been the technology capability at the time to handle thousands of users? What would the military and security consequences have been? What was the capability of enemy signal intercept? Some people may want to disparage the FCC, but what was the rationale for the decision to be so limiting? This next section attempts to answer these questions, and more.

9.3.1 Communications Act of 1934: The Beginning of a 30-Year Mobile Drought

By 1934, the need had arisen for consolidation and control of the communications systems that were being built in the United States. This act effectively both created the Federal Communication Commission and subjugated the entire telecom industry to regulatory compliance by combining existing laws pertaining to both radio and telephone communications into one body, and was signed into law by President Franklin Roosevelt in June 1934. Jurisdictional foundation for federal regulation of the systems derived primarily from the precedence of governance regarding interstate commerce, the sole authority over which resides in the federal government.

On the surface, the reason for the act, and the reason that it purported:

> Regulating interstate and foreign commerce in communication by wire and radio so as to make available, so far as possible, to all the people of the United States a rapid, efficient, nationwide, and worldwide wire and radio communication service with adequate facilities at reasonable charges, for the purpose of the national defense, and for the purpose of securing a more effective execution of this policy by centralizing authority theretofore granted by law to several agencies and by granting additional authority with respect to interstate and foreign commerce in wire and radio communication, there is hereby created a commission to be known as the 'Federal Communications Commission', which shall be constituted as hereinafter provided, and which shall execute and enforce the provisions of this Act.

It seems to be logical and would surely only introduce great efficiencies. Behind the scenes, however, and just by scratching at the surface, evidence suggests collusion between certain telecom companies, their lobbyists, and key politicians. In a Wikipedia article on this subject, the history of the motivations behind this act is relegated to just a single line, which attributes the formation of the FCC and subsequent regulations to the lobbying efforts of the National Association of Regulatory Utility Commissioners (NARUC)—today NARUC is a very large organization that includes state-level commissioners from nearly every aspect of publicly regulated utilities, of which the communications companies are an aspect.

To truly understand the origin of the FCC and its creation, we look first to the creation of the Federal Radio Commission (FRC) in 1924.* Three primary players that held key positions in government were at the vortex of bringing order out of the perceived industry chaos of the early years of radio.

Imagine the Internet in its nascence, only without the traffic directing capabilities and with no agency to govern the use of domain names. Imagine, sitting in your home, that every time you went to scorellis.net, you could get one of any four websites. Not only that, but half the images on the page would be from one of any four of the websites. It would be a mess. In the early days of radio, broadcasts could, and did, step all over each other. The programming that would reach listeners would boil down to who had the greatest signal strength, and Hoover would use this fact to promote the establishing of regulations that would prevent broadcast interference. So it was in this climate that Secretary of Commerce Herbert Hoover, Maine Representative Wallace White, and Washington Senator Herbert Dill embarked, not necessarily together, on a campaign to establish a unifying Federal Radio Commission that would control the airwaves, the establishment of which would have far-reaching implications for the next 75 years, and would delay and hamper development of mobile technologies until the late 1960s, when the FCC ultimately reversed its position and began a deregulation initiative that would, moving at the speed of government, take 30 years to complete.

9.3.2 Herbert Hoover

Secretary of Commerce Herbert Hoover understood that the radio industry was practically begging to be regulated. The key players in the industry

---

* "The Ideological Fight over the Creation of the Federal Radio Commission in 1927," *Journalism History*, vol. 26, no. 3, 2000.

understood how easily anyone with a radio transmitter could begin transmitting and broadcasting their own content, and he also knew that it could be subversive content of nefarious intent. He could not have envisioned or imagined a digitally enabled system that would allow for many thousands the number of broadcast channels, nor the positive, progressive impact that leaving the airwaves free would have engendered.

Not until the mid-1990s, when the Internet enabled anyone with a website to broadcast their message, would the benefits of this freedom (along with its darker elements) be realized. Ultimately, he believed that the potential chaos that would ensue if the airwaves remained unregulated would result in anarchy and a system over which the government had no control. He wanted a "radio kill switch"—that is, he believed that the government, during a time of emergency, should have complete and unfettered access to the airwaves.

It's easy to see how this philosophy remained entrenched in the security policy of government security agencies and ultimately led to the breach of millions of citizens' rights to privacy, which are assured under the constitution, but on the publicly controlled "information super highway" to which the mobile networks are so tightly integrated today, the policies of Hoover become readily enforceable. He played a key, pivotal role in the creation of the FRC.

9.3.3 Wallace White

Maine Representative Wallace White, a prominent Republican who held office from 1916 to 1929 as a representative, and then from 1930 to 1949 in the Senate, attended many international radio and telegraph conferences as a delegate. He is credited on the official U.S. Senate history website with overcoming the "chaos" of the airwaves as a coauthor of the Radio Act of 1927. He worked closely with Hoover to develop and author the act. Hoover, through his own manipulations, had actually worked to *increase* and *accelerate* the chaos.

White attended Hoover's radio conferences, and ultimately forged a relationship with the Radio Corporation of America (RCA) that would be at the heart of the subsequent legislation—a relationship that progressives of the time would fear as being designed to set RCA up to become a monopoly.

9.3.4 Herbert Dill

Washington Senator Herbert Dill sided with the progressives and provided the role of antagonist during the creation of this legislation, and submitted competing legislation. Throughout the negotiations for the new bill, both White and Hoover would meet with RCA executives and legal staff, and ultimately represented the interests of the radio industry. In a telegram to White, an RCA executive states:

Meet for lunch Tuesday at 1 with SARNOFF [sic]

David Sarnoff, the son of immigrant Belarusians, had worked his way up through the Marconi Wireless Telegraph Company of America, which was later purchased by Owen Young of General Electric and renamed the RCA. Subsequently, and perhaps not coincidental to meetings with certain government representatives that would work to ensure favorable market conditions, RCA purchased its first radio station in 1926, WEAF of New York, and established the National Broadcasting Company (NBC). It would be the first radio network to be established in America. White would also meet with Lloyd Espenschied of AT&T, who also happened to be a majority shareholder in NBC. Espenschied, an inventor, possibly advised White in an expert capacity. Part of the defensibility of the bill was that the development of the system of telecom would be built and developed by experts, one of whom no doubt was Espenschied.

It becomes apparent, then, that the nascence of the FCC is deeply rooted in the interactions that ultimately led to the creation of the FRC by White and Hoover. Dill, while influential, held little power in the creation of the bill. White was a genius of bipartisanship—he was unassailable by either party and, while not known to be a great statesman, was considered a brilliant author of legislation. Dill's proposed legislation would result in a vehement rebuttal by Hoover, written in the margins of the bill, mostly to the affect of it being "useless, unfair bunk." One of these two men would later become president of the United States of America.

9.3.5 The Argument

Dill believed that the newly formed commission should be free of politics and industry influence, should be independent radio, and should be the instrument, created by the people and controlled by the people. Hoover (and by proxy, White) believed that people lacked the fundamental ability to determine and make proper content choices for themselves. They both agreed that government should be in control. At the heart of the argument, the primary difference between what Hoover wanted and what Dill wanted lay in one aspect: prevention of monopoly. Dill was antimonopoly, and White and Hoover wanted to create an agency that would, through the strict regulation of radio broadcast technology through just a few state-controlled/cooperating agencies (GE, RCA, AT&T, NBC), engender, support, nurture, and enable the creation of a monopoly that would subsequently control radio-based transmissions for the next 70 years.

9.3.6 The Agenda

Behind Hoover's initiative lay a publicly disclosed agenda. These items would serve as the core tenants of the regulation and would not see reprieve until the Internet broke all boundaries and unleashed a tidal wave of information and a platform from which to present them.

Protect the public: The government believed that the people needed to be protected from harmful, thought-provoking speech that could lead to incendiary speech. Only responsible parties (American broadcasting corporations) that possessed revenue-generating relationships (advertising) could reap the profits and revenue needed to build out the infrastructure in an organized fashion that served the public. In a speech to the participants of a radio conference in 1924, Hoover declared that radio broadcasters would maintain free speech, and that the government would ensure this, but that they would also maintain them (the radio waves) free of malice and unwholesomeness.

Ensure growth through technical regulation: This dovetails nicely into the previous requirement. By regulating the equipment that could be used, and dictating broadcast ranges, frequencies, and through licensing, signal interference could be prevented and the quality of transmissions would be assured.

Create a more efficient government: Government is in the business of governing people, and to enact its governance, what better way than to possess the ability to send messages into their homes? The weekly president's radio address still reaches millions, and we can all subscribe to text messages from our senators and representatives as well as participate in town-hall-style meetings via our cellular phones. The Executive Branch, the Treasury, the Department of War, the Department of Justice, the U.S. Post Office, and the Navy were just some of the agencies that would be identified as benefiting from regulatory control of the airwaves.

Serving in the public interest: In the absence of technology to control the receiving of transmissions, ostensibly one could be subjected to listening to the transmissions of anyone that chooses to broadcast. Imagine a Facebook message thread where any of your friends' friends can post ideas and thoughts to the thread for you to see, only this sort of thing happens on your radio, and you don't even have control over it. At least on Facebook you can block, hide, or "unfriend" offensive people. No such capability existed in the 1920s. But had we not regulated it, and had forced an environment (not unlike the Internet) where innovation would be the only way whereby broadcasting could be controlled, through technical invention, would such invention have developed? We can't know, but it's entirely possible that such technology was already being developed; often, when a need is prevalent enough, and enough people see it, an invention emerges that solves the problem. In the meantime, the only option for people would be to turn off their radio. The risk that people would in fact do just that frightened everyone into submission to the idea of a government-regulated industry.

9.3.7 Regulation Emerges

In 1926, Dill, White, and Hoover managed to come to a sort of agreement that created the FRC—essentially, a temporary organization. Ultimately, less than 10 years later, President Hoover would consolidate the power of control over radio into a single organization, the FCC, which exists today, albeit with little or no control over the content distributed on the Internet.

9.3.8 1968: The Beginning of 30 Years of Reform

In 1968, due to attempts by AT&T to maintain its vice-like grip against other devices attaching to its network, a mobile phone company, Carterfone, brought a suit against certain Bell companies. Thomas Carter had invented a wireless device that enabled two-way communications between devices. It used the telephone network as a backbone.

The following text comes from the actual decision. The only changes that have been made to it have been formatting, and I have emboldened what I think are the most interesting and relevant parts, and italicized the parts that are even more interesting than that:

In the Matter of USE OF THE CARTERFONE DEVICE IN MESSAGE TOLL TELEPHONE SERVICE; In the Matter of THOMAS F. CARTER AND CARTER ELECTRONICS CORP., DALLAS, TEX. (COMPLAINANTS), v. AMERICAN TELEPHONE AND TELEGRAPH CO., ASSOCIATED BELL SYSTEM COMPANIES, SOUTHWESTERN BELL TELEPHONE CO., AND GENERAL TELEPHONE CO. OF THE SOUTHWEST (DEFENDANTS)

Docket No. 16942; Docket No. 17073
FEDERAL COMMUNICATIONS COMMISSION
13 F.C.C. 2d 420 (1968); 13 Rad. Reg. 2d (P & F) 597
RELEASE-NUMBER: FCC 68-661
June 26, 1968 Adopted

On December 21, 1966, Carter filed a formal complaint pursuant to section 208 of the Communications Act, 47 U.S.C. § 208, against General and certain of the Bell companies, and further proceedings in docket No. 16942 were held in abeyance pending disposition of the complaint (docket No. 17073). By order released March 8, 1967, the complaint was consolidated for hearing with docket No. 16942, and the following issues were added:

1. Whether, with respect to the period from February 6, 1957, to December 21, 1966, the regulations and practices in tariff FCC No. 132 of the American Telephone and Telegraph Co. were properly construed and applied to prohibit any telephone user from attaching the Carterfone device to the facilities of the telephone companies for use in connection with interstate and foreign message toll telephone service; and if so
[*423]2. Whether, during the aforesaid period, such regulations and practices were unjust and unreasonable, and therefore unlawful within the meaning of section 201(b) of the Communications Act of 1934, as amended, or were unduly discriminatory or preferential in violation of section 202(a) of said Act.

The examiner found that there was a need and demand for a device to connect the telephone landline system with mobile radio systems which could be met in part by the Carterfone. He also *found that the Carterfone had no material adverse effect upon use of the telephone system.* He construed the tariff to prohibit attachment of the Carterfone whether or not it harmed the telephone system, and determined that future prohibition of its use would be unjust and unreasonable. He also found that it would be unduly discriminatory under section 202(a) of the Act, since the telephone companies permit the use of their own interconnecting devices. However, he did not find the tariff prohibitions to have been unlawful in the past, largely because the harmless nature of the Carterfone was not known to the telephone companies, and he did not find that a general prohibition against nontelephone company supplied interconnecting devices was unjust or unwise, because of the risk he saw of "serious harm to the heart of the nation's communications network."

We agree with and adopt the examiner's findings that the Carterfone fils [sic] a need and that it does not adversely affect the telephone system. They are fully supported by the record. We also agree that the tariff broadly prohibits the use of interconnection devices, including the Carterfone. Its provisions are clear as to this. **Finally, in view of the above findings, we hold, as did the examiner, that application of the tariff to bar the Carterfone in the future would be unreasonable and unduly discriminatory. However, for the reasons to be given,** *we also conclude that the tariff has been unreasonable, discriminatory, and unlawful in the past, and that the provisions prohibiting the use of customer-provided interconnecting devices should accordingly be stricken.*

We hold that the tariff is unreasonable in that it prohibits the use of interconnecting devices which do not adversely affect the telephone system. See Hush-A-Phone Corp. v. U.S., 99 U.S. App. D.C. 190, 193, 238 F. 2d 266, 269 (D.C. Cir., 1956), holding that a tariff prohibition of a customer supplied "foreign attachment" was "in unwarranted interference with the telephone subscriber's right reasonably to use his telephone in ways which are privately beneficial without being publicly detrimental." n2 The principle of Hush-A-Phone is directly applicable here, there being no material distinction between a foreign attachment such as the Hush-A-Phone and an interconnection device [*424] such as the Carterfone, so far as the present problem is concerned.

n3 Even if not compelled by the Hush-A-Phone decision, *our conclusion here is that a customer desiring to use an interconnecting device to improve the utility to him of both the telephone system and a private radio system should be able to do so, so long as the interconnection does not adversely affect the telephone company's operations or the telephone system's utility for others.* A tariff which prevents this is unreasonable; it is also unduly discriminatory when, as here, the telephone company's own interconnecting equipment is approved for use. The vice of the present tariff, here as in Hush-A-Phone, is that it prohibits the use of harmless as well as harmful devices.

n2 After Hush-A-Phone, the Commission directed A.T. & T. to "file tariff schedules * * * rescinding and canceling any tariff regulations to the extent that they prohibit a customer from using, in connection with interstate, or foreign telephone service, the Hush-A-Phone device or any other device which does not injure defendants' employees, facilities, the public in its use of defendants' services or impair the operation of the telephone system." Hush-A-Phone, decision and order on remand, 22 F.C.C. 112 (Feb. 6, 1957).

The Commission additionally stated in its decision and order on remand: "As we construe the court's opinion, a tariff regulation which amounts to a blanket prohibition upon the customer's use of any and all devices without discriminating between the harmful and harmless encroaches upon the right of the user to make reasonable use of the facilities furnished by the defendants."

The modification of the offending tariff provision filed by A.T. & T., and designated paragraph B24 of tariff FCC No. 132, is at issue here.

n3 The Hush-A-Phone was a cup-like device mechanically fastened to the mouthpiece of a telephone handset. **The Carterfone by means of acoustic and inductive coupling effectively achieves an "interconnection" between the public toll telephone system and private mobile radio systems. These differences are immaterial, however, insofar as the Hush-A-Phone holding is concerned.**

A.T. & T. has urged that since the telephone companies have the responsibility to establish, operate and improve the telephone system, they must have absolute control over the quality, installation, and maintenance of all parts of the system in order effectively to carry out that responsibility. Installation of unauthorized equipment, according to the telephone companies, would have at least two negative results. First, it would divide the responsibility for assuring that each part of the system is able to function effectively and, second, it would retard development of

the system since the independent equipment supplier would tend to resist changes which would render his equipment obsolete.

There has been no adequate showing that nonharmful intercon-nection must be prohibited in order to permit the telephone company to carry out its system responsibilities. The risk feared by the exam-iner has not been demonstrated to be substantial, and no reason presents itself why it should be. No one entity need provide all interconnection equipment for our telephone system any more than a single source is needed to supply the parts for a space probe.

We are not holding that the telephone companies may not prevent the use of devices which actually cause harm, or that they may not set up reasonable standards to be met by interconnection devices. These remedies are appropriate; we believe they are also adequate to fully protect the system.

This decision opened up remarkable opportunities for the development of the cellular phone technology, and ultimately also enabled the Internet. Imagine a world today where only AT&T-provided equipment could be connected to the communications backbone. Free marketplaces are essential to innovation; rigidity and heavy regulation stymie free thought and creativity.

9.3.9 Deregulation, the Internet, and Mobile Internet

Since so much of this book, and this chapter, revolves around the history of the telephone, Jason Oxman[*] asserts that while the FCC certainly can-not take responsibility for the creation of the Internet, it did, through its policy of deregulation, which began in 1968 with the Carterfone decision, provide the fertile, unregulated environment needed to spur growth and development of it. In 1966, the FCC had commenced the first formal com-puter inquiry into the possibilities inherent in the integration of comput-ers and telecommunications.[†]

The end result is, arguably, the beginning of a pattern of deregula-tion that would take nearly 30 years to implement. Again, the pattern of 50–60 years of technology growth is evidenced.

Before 1965, the relationship between the FCC and the phone monopo-lies was incestuously collaborative. As a regulated monopoly, Ma Bell and

---

[*] Jason Oxman, "The FCC and the Unregulation of the Internet," OPP Working Paper 31, 1999.

[†] "In the Matter of Regulatory and Policy Problems," presented by the Interdependence of Computer and Communication Services and Facilities, 7 FCC 2d 11, 1966 (first computer inquiry).

Western Electric purportedly acted in, from their perspective, the best interests of the public.

Ultimately, what this really meant was that they operated in the best interests of themselves, which meant that the government afforded protections to the monopolies, shielding them from competitive incursions, while at the same time protecting them from the possibility of government ownership.

By late 1966, things began to change and the atmosphere began to take on an ambience of true public ownership of the communications system. In particular, the FCC began to understand that a new age of computing was dawning, and that in order to remain both competitive and technologically on par with other countries, a stale and antiquated phone system could not be tolerated. Primarily, on the surface, fostering of the free market would provide a more technically advanced telecom that would enhance commerce and bolster trade, but one cannot ignore the military applications of improved communication systems as well. Time and time again, throughout history, the power and capability of communication systems surfaces as one of the greatest success factors in military activities, namely, war.

9.4 WEAPONIZATION OF MOBILE TECHNOLOGY

Since time immemorial, the efficiency of communications has played a massive role in the execution of battle. The ability to control the deployment of forces, to respond to changing battlefield conditions, depends largely on the ability to collect information and rapidly (and effectively) make decisions about the data that will result in a battlefield advantage.

Drums, signal fires, smoke, flags, mirrors, runners, horseback, and flares were the earliest forms of signal communications. Wired communications introduced a communication that could easily be disrupted and intercepted. Enemy troops could easily cut telegraph lines or tap in to them to intercept communications. With this capability the "signal corps" emerged—a branch of military discipline that specialized in three areas:

- Interception of communication

- Disruption of enemy communications

- Securing and identification of friendly communications

The limitless military application of wireless communication had become very apparent by 1917 when, on April 7, the U.S. government, elbows deep in WWI, banned all radio communications within the United States. In fact, the entire nation fell radio silent.

The year 1917[*] saw published research that outlined research efforts by AT&T that included two-way voice ground-to-air communication airplanes that would ultimately allow for air formation management, just as infantry units on the ground.

Flash forward to 2014: military members are testing cell phones for use in battlefield conditions, using them to send field reports.

9.4.1 The Future of Net Neutrality

1946: First Mobile Telephone Call[†]

June 17, 1946—A driver in St. Louis, Mo., pulled out a handset from under his car's dashboard, placed a phone call and made history. It was the first mobile telephone call.

A team including Alton Dickieson and D. Mitchell from Bell Labs and future AT&T CEO H.I. Romnes, worked more than a decade to achieve this feat. By 1948, wireless telephone service was available in almost 100 cities and highway corridors. Customers included utilities, truck fleet operators and reporters. However, with only 5,000 customers making 30,000 weekly calls, the service was far from commonplace.

That "primitive" wireless network could not handle large call volumes. A single transmitter on a central tower provided a handful of channels for an entire metropolitan area. Between one and eight receiver towers handled the call return signals. At most, three subscribers could make calls at one time in any city. It was, in effect, a massive party line, where subscribers would have to listen first for someone else on the line before making a call.

Expensive and far from "mobile", the service cost $15 per month, plus 30 to 40 cents per local call, and the equipment weighed 80 pounds. Just as they would use a CB microphone, users depressed a button on the handset to talk and released it to listen.

Improved technology after 1965 brought a few more channels, customer dialing and eliminated the cumbersome handset. But

---

[*] *Journal of Electricity*, July 15, 1917, p. 59.
[†] See http://www.corp.att.com/attlabs/reputation/timeline/46mobile.html.

capacity remained so limited that Bell System officials rationed the service to 40,000 subscribers guided by agreements with state regulatory agencies. For example, 2,000 subscribers in New York City shared just 12 channels, and typically waited 30 minutes to place a call. It was wireless, but with "strings" attached.

9.4.2 The Cellular Solution

Something better—cellular telephone service—had been conceived in 1947 by D.H. Ring at Bell Labs, but the idea was not ready for prime time. The system comprised multiple low-power transmitters spread throughout a city in a hexagonal grid, with automatic call handoff from one hexagon to another and reuse of frequencies within a city. The technology to implement it didn't exist, and the frequencies needed were not available. The cellular concept lay fallow until the 1960s, when Richard Frenkiel and Joel Engel of Bell Labs applied computers and electronics to make it work.

AT&T turned its work into a proposal to the Federal Communications Commission (FCC) in December 1971. After years of hearings, the FCC approved the overall concept, but licensed two competing systems in each city.

In 1978, AT&T conducted FCC-authorized field trials in Chicago and Newark, New Jersey. Four years later, the FCC granted commercial licenses to an AT&T subsidiary, Advanced Mobile Phone Service (AMPS). AMPS was then divided among the local companies as part of the planning for divestiture. Illinois Bell opened the first commercial cellular system in October 1983. AT&T reentered the cellular business by acquiring McCaw Cellular in 1994, the same year that President Clinton awarded Frenkiel and Engel the National Medal of Technology.

Today, AT&T Wireless operates one of the largest digital wireless networks in North America. With more than 17 million subscribers, including partnerships and affiliates, and revenues exceeding $10 billion, AT&T Wireless is committed to being among the first to deliver the next generation of wireless products and services. AT&T Wireless offers high-quality wireless communications services, whether mobile or fixed, voice or data, to businesses and consumers in the United States and internationally.

CHAPTER 10

Cellular Access Systems

Lauren Collins and Scott R. Ellis

CONTENTS

10.1 DELINEATION

In the United States, for the most part, there are two types of cellular networks used: CDMA2000 and GSM. It gets confusing because there are so many acronyms used together. One way to think of the difference between GSM and CDMA2000 is like the difference between gasoline and diesel. Your car probably says "unleaded" on it. GSM is a *standard*, and if something is compliant to that standard, then the equipment will be interoperable. A GSM phone, then, will not interoperate with CDMA2000. Similarly, car and equipment manufacturers make phones and equipment that meet both standards. For example, you can get an iPhone that will work on a GSM or a CDMA2000 network, but the same phone will never work on both—the equipment is different. Long Term Evolution (LTE) is a relatively new standard poised to replace GSM and CDMA2000, but for the sake of simplicity, this chapter focuses on the most prevalent, current technology and assumes that you can swap the terms later, once you are comfortable. Let's walk before we run, and avoid falling and skinning our mental knees. Ultimately, someday, everything may be LTE, but since foreign countries have different radio band restrictions, only multiband phones will work.

This chapter delineates the differences between two types of networks and discusses the range, distribution, and on a surface level of detail, the hardware and technology used to build out these networks. In the United States, the term *cell phone* is the most prevalent moniker for mobile handset phones. Other countries use the term *mobile*, which is perhaps a more accurate description of the phone itself. In this chapter, the word *mobile* is used to describe the handheld transceiver that moves in and out of cellular-based transceivers. The understanding, then, is that the term *cell phone* is an American term, because it is clear that the actual cell phone is, in fact, the equipment that sits in a somewhat stationary tower.

Cellular technology has the unique requirement that it must be able to handle many simultaneous connections by many unknown devices. The burden of authentication, data transfer, and connection management resides on the cell tower. There are several groupings of technical equipment interfaces that come together to make this happen.

Cellular communications operate on a frequency that is similar in range to that of your standard, cordless home phone handset. In the last few years, the lines have blurred somewhat, but there are still some pretty big differences. The Federal Communications Commission (FCC) has allocated the following frequencies for home handsets:

- 1.7 MHz

- 43–50 MHz

- 900 MHz (902–928 MHz)

- 1.9 GHz (1880–1900 MHz)

- 1.9 GHz (1920–1930 MHz)

- 2.4 GHz

- 5.8 GHz

Alternatively, cellular-based mobile devices operate in a series of overlapping, often hexagonal cells. The frequencies aren't actually hexagonally broadcast—that would be impossible. Instead, the geography is divided into hexagon-shaped areas of land, as demonstrated in Figure 10.1.

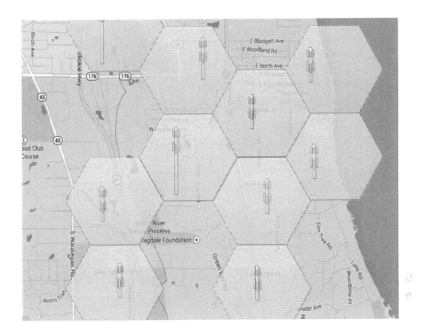

FIGURE 10.1   Maps are carved up by geometry, with cell sites placed at or near the center of each hexagon. Certain structures may be eligible to lease towers: water towers, tall buildings, existing fire and police communications antennas, and warning siren poles.

Hexagons fit together rather nicely, and resemble circles, which do not line up or look nice on a map. Each cell has its own dedicated range, and there are some simple math formulas that dictate the distance between two cell towers that is necessary for them to be allowed to use the same signal. Clearly, in order to have clear transmissions, adjacent cell towers cannot use the same frequencies. The frequencies allocated to cell towers by the FCC fall into the ultra-high-frequency (UHF) range. Different countries have different ranges that they use, but they all fall into the same UHF class.

Frequency bands for sites are divided into bands, and are sold geographically in large "trading areas" by the FCC. When frequencies become available, they are auctioned and sold to the highest bidder. The winner may use the spectrum for any purpose. The United States 700 MHz FCC Wireless Spectrum Auction of 2008, for example, was won by Verizon and AT&T. Formerly, 700 MHz was the spectrum used by analog television broadcasting.

10.2 CELLULAR ACCESS TECHNOLOGIES: FDMA, TDMA, CDMA, AND SDMA

This book has made an effort to simplify the alphabet soup into just a few terms that can be paired as follows:

- GSM pairs with frequency-division multiple access (FDMA) and time-division multiple access (TDMA) (between cells and within cells).

- CDMA2000 pairs with code-division multiple access (CDMA).

- LTE pairs with OFDM (but remember, we aren't talking about LTE).

- Spatial-division multiple access (SDMA) is in a league of its own (you'll see why).

GSM providers (AT&T) and CDMA2000 providers (Verizon) use TDMA and CDMA, respectively. At the tower equipment level, then, the difference between the two is the type of radio equipment that is used (three, actually, but we promised you earlier not to talk about LTE). Some providers are also using FDMA and LTE, and still others are creating confusion in the marketplace by referring to LTE as 4G. FDMA provides different frequency bands to different data streams, allocating the data streams to different nodes or devices. An advanced form of FDMA known as orthogonal frequency-division multiple access (OFDMA) is used in 4G cellular communication. Table 10.1 differentiates the standards from the generations of mobile technology systems.

Each has its pros and cons, and extensive research does not indicate that one is superior to the other. GSM has achieved a higher degree of global acceptance, with the most market share.

In GSM, the communications are handled between cells by dividing the frequency so that adjacent towers do not overlap. Within the cell, time division is used, which was described at length in Chapter 9. You may think of TDMA (referred to in this book as "too darn many acronyms" (TDMA)) as a chat room, where the messages are scrolling upward, one at a time, and people know whose conversation they are following. Eventually, though, when under load, the scrolls' speed increases and you can no longer keep up. There are only a set number of conversations that can be viewed.

CDMA is more like a monitor screen, with many messages popping in simultaneously. The message that you want to see is a certain color, and so

TABLE 10.1 Mobile Generations
and Standards from the Past,
Present, and Also 5G for the Future

Generation	Standard
2.5G	GPRS
2.75G	EDGE
3G	UMTS
	W-CDMA
	HSPA 3.6
	HSPA 7.2
Pre-4G	HSPA 14
	HSPA+
	WiMAX
	LTE
4G	WiMAX2
	LTE-Advanced
5G	TBD

you recognize your conversation by its color. Eventually, though, there are so many messages popping in that they keep getting smaller and smaller, and it becomes difficult to read your message without getting closer. The more messages there are, the harder it is to distinguish your color from other colors that are near it, and the messages get smaller, and harder to read, so you have to get closer to the monitor (the cell site) to be able to see them and to pick out your frequency (your color).

Interestingly, the real-world counterparts to these examples begin to break down in similar ways, too. As CDMA networks begin to get overloaded, the range of the tower begins to shrink, which explains why you might have great signal at a location one day, and terrible the next. Perhaps you live next to a ballpark and when there is a game, your signal gets depleted. With GSM, there are simply a set number of connections that can be held, and users are subjected to dropped calls.

SDMA reuses the same set of cell frequencies in a given service area. Through the use of smart antennas with beams directed at the mobile station, SDMA serves disparate users within the same region. SDMA increases the capacity of the system and transmission quality by focusing the signal into narrow transmission beams. Any mobile stations operating outside of the bounds of said directed beams experience near-zero interference from other mobile stations that operate under the same base station with the same frequency.[*] And because the beams are focused, the radio energy frequency has increased base station range. This is an attribute of SDMA that permits base stations to have larger radio coverage with less radiated energy. Thus, the narrow beam width also sanctions more gain and clarity.

Base stations in legacy cellular network systems radiate radio signals in all directions within the cell, having no knowledge of the location of a mobile station. Subsequently, SDMA technology channels radio signals constructed on the location of the mobile station. Utilizing this method, SDMA architecture saves valuable network resources and prevents redundant signal transmission in areas where mobile devices are currently inactive. A distinct advantage of SDMA is frequency reuse. Provided the reuse distance is preserved in network architecture, interference is near zero, even if mobile stations utilize the same allocated frequencies.

Table 10.1 lists 5G at the bottom as TBD (to be determined). Multiple projects are in play for the research and development of 5G, and the consensus is for 5G to deliver faster data with more energy efficiency, while simultaneously including advancements. As ubiquitous computing becomes a more popular goal each day, networks that provide Internet of things and wireless sensor networks would more seamlessly move data between the data paths with the access technology proposed for 5G.[†] Another driver of 5G technology is Li-Fi, a hybrid of light and Wi-Fi using light-emitting diodes to transmit data, rather than utilizing radio waves, as used for Wi-Fi.[‡]

10.3 CELLULAR ACCESS ARCHITECTURE

Traditional cellular systems perished due to poor functionality between radio base stations and base station controller equipment, as they were utilizing time-division multiplexing (TDM) or asynchronous transfer mode (ATM). The cellular transport networks were forced to evolve to meet the demands for increased data needs.

---

[*] Cory Janssen, "What Is Spatial Division Multiple Access (SDMA)?—Definition from Techopedia," Techopedias, n.d., http://www.techopedia.com/definition/2979/spatial-division-multiple-access-sdma (accessed June 15, 2014).

[†] Abdullah Gani, Xichun Li, Lina Yang, Omar Zakaria, and Nor Badrul Anuar, "Multi-Bandwidth Data Path Design for 5G Wireless Mobile Internets," *WSEAS Transactions on Information Science and Applications Archive*, 6(2), 2009.

[‡] National Instruments and the University of Edinburgh Collaborate on Massive MIMO Visible Light Communication Networks to Advance 5G, Cambridge Wireless, November 20, 2014.

FIGURE 10.2   A 2014 photograph of a cell tower that rises above the Lake Bluff fire station. (Photograph courtesy of Scorellis Productions.)

10.3.1 Anatomy of a Cell Tower

As the name *cellular* implies, the eponymously named networks are geographically wide area networks built of heavy-duty, tower-based transmitters that are accessed by mobile transponders. The image in Figure 10.2 is of a cellular tower that rises above small-town suburb Lake Bluff, Illinois.

The main components of cellular towers are as follows:

- Base transceiver station (BTS): Radio transceiver equipment that communicates with mobile phones.

- Base station controller (BSC): A controller to manage the transceiver equipment and channel assignments.

- Mobile switching center (MSC): The cellular network switches.

10.3.2 Base Transceiver Station (BTS)

The towers and the equipment installed to them that you see on buildings, towers, and water towers all around you are, essentially, the BTS. The following equipment goes into its construction:

- Transceiver (TRX): The transceiver, or the driver receiver (DRX), can be comprised of a single radio unit (sTRU), a double radio unit (dTRU), or a composite radio unit. It handles transmission and reception of signals, as well as handling communications to the tower controller (BSC).

- Power amplifier (PA): This unit amplifies the signal received by the TRX.

- Duplexer: The duplexer provides duplexing capabilities to the antenna so that signals can be both sent and received with just one antenna.

- Antenna: These are the oblong plates that you see attached to cell towers and are in fact shrouds that encase high-powered microwave antenna.

- Alarm extension system: This is the equipment that collects fault monitoring for various pieces of equipment with the BTS, and sends them to the monitoring stations.

- Baseband receiver unit (BBxx): All modern signal processing algorithms are embodied in the electronics of this unit. Here is where the FDMA and CDMA techniques allow for crowd voice handling and handoff between towers. Less busy towers may have signals pushed to them by busy towers.

The photo illustrated in Figure 10.3 is a close-up of cellular tower components. In the simplest terms, the cell tower equipment follows the same paradigm as your home system, should you have a wireless phone in your house. Your wireless home system has a handset, a base station, and then the Internet router to which your base connects via a hardwire, and a central office that routes the call.

10.3.3 Network System

It seems incredible that so many cell phones can be on one network. For example, at busy sports events, parades, etc., how can the infrastructure be sure to bear the load? The answer is: sometimes, it can't. Each tower is connected to a junction box, which acts as a server to take your call and route it over fiber to, well, whoever you want. That might be a web page, an email, or a voice call that you are making to your neighbor. Like any infrastructure system, if you want more users, you need more bandwidth, and more electronics to handle the communications. Future scalability will require massive parallel systems. Every household will have a direct fiber link, and smaller cells will be the order of the day, with radii that are

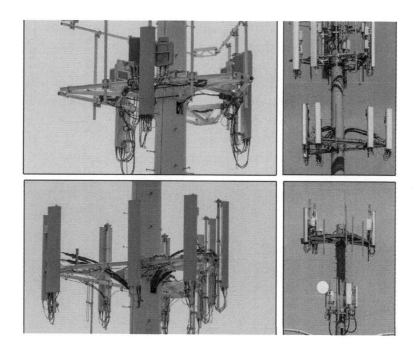

FIGURE 10.3 These photographs present several common site configurations, including a directional, round antenna that is aimed at the town firehouse (bottom right).

predetermined by population and demand, with mini-cells firing up as needed and providing service everywhere.

Figure 10.4 illustrates a legacy network architecture, where the base transceiver station (BTS) incorporates all the radio equipment that support the physical layer channels. The scope of radio equipment functionality included modulating and demodulating signals to and from the mobile

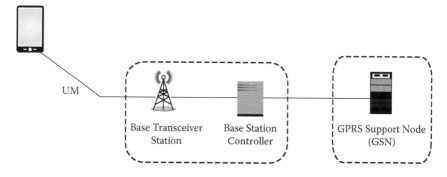

FIGURE 10.4 Legacy base station and network switching subsystem architecture provide simple connectivity supporting a limited feature set and limiting speeds.

device over the universal mobile (UM) interface, as well as channel coding and encryption. Considering the applications running on mobile devices today, you may already realize the need for an architecture makeover. The base station controllers (BSCs) handle hundreds of base transceiver stations (BTSs). Since a BSC is capable of handling some switching, it will handle control and bearer plane traffic to and from the BTSs, as well as operations and maintenance messages. In addition, the switching capability is also able to provide the handoff between neighboring BTSs as a mobile device moves from one cell tower to another. Channel configurations can be controlled by one BSC even as a device transitions to a neighbor BTS.

To put things into perspective, the model illustrated in Figure 10.4 (and described above) was only capable of sending 140-bit SMS messages at an aggregate throughput of less than 1500 bps.* Data services could no longer be handled efficiently as the volumes of traffic surpassed legacy circuit-switched voice and mobile access systems. Today, edge networks are implemented solely for the purpose of optimizing the scheduling of packet delivery with efficient IP service delivery in mind.

---

* Juha-Pekka Ahopelto, Juha-Pekka, and Hannu Kari, "Packet Radio System and Methods for a Protocol-Independent Routing of a Data Packet in Packet Radio Networks," Patent US5970059, Nokia Telecommunications Oy, January 9, 1986.

CHAPTER 11

Wireless Access Systems

Lauren Collins

CONTENTS

IP TECHNOLOGY IS A PROGRESSIVE COMPONENT of mobile networking that permits end-to-end solutions to handle data efficiently. Subsequently, data growth must be handled accordingly as bit rate increases. By limiting the number of nodes, latency is low and real-time services over the packet domain make everyone happy. Figure 11.1 illustrates a present-day mobile switching architecture. This chapter takes the reader through various wireless deployments and makes an effort to appreciate the technology we use today.

11.1 WLAN

There was a point in time where home and business owners were able to purchase and implement wireless technologies that outperformed cellular services. The answer was not to roll out wireless local area networks

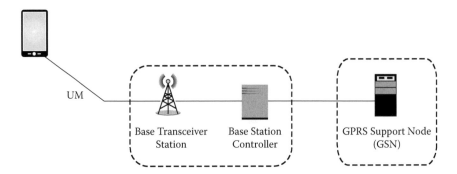

FIGURE 11.1 Legacy base station and network switching subsystem architecture provide simple connectivity supporting a limited feature set and limiting speeds.

(WLANs) in an effort to replace the cellular carriers and infrastructure; these issues set the standards for cellular and mobile systems to provide the ability to internetwork WLAN technologies with voice services. Even today, in-building wireless infrastructures would resolve the need to implement WLAN in metropolitan office buildings. When IEEE approved 802.11g as a standard in 2003, bit rates of up to 54 Mbps were offered. The comparable cellular services provided connectivity of only 40 kbps. Today with 802.11ac, wave 1 presents a speed of 7 Gbps in the 5 GHz band.* Although most engineers want the latest and greatest and fastest, the speed associated with 802.11ac presents a massive increase in traffic on corporate networks, making data capture and traffic analysis nearly impossible.

WLAN systems used unencrypted communications that are vulnerable to attacks, so the world was not too quick to deploy mass amounts of these infrastructures. Once IEEE standardized port-based authentication, MAC security enhancements followed, and the Internet Engineering Task Force's (IETF) Extensible Authentication Protocol (EAP) provided a secure public WLAN. Engineers were able to incorporate two services into their architecture, Direct IP Access and 3G/LTE IP Access. Figure 11.2 displays Direct IP Access, which provides a WLAN device IP connectivity to the LAN, and 3G/LTE IP Access, which employs Packet Data Gateway (PDG) to allow the WLAN device access to services relevant to cellular services.

Direct IP Access authentication gets transported over authentication, authorization, and accounting (AAA) interfaces, having 3G/LTE

---

* News & Events: Press Releases, "New IEEE 802.11ac™ Specification Driven by Evolving Market Need for Higher, Multi-User Throughput in Wireless LANs," n.p., January 7, 2014.

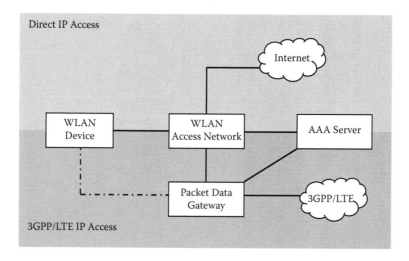

FIGURE 11.2   Direct IP Access and LTE IP Access allow the WLAN device to access cellular services as well as WLAN secure services.

delineating both RADIUS and diameter protocol options. When accessing 3G/LTE distinctive services, including corporate or home-based services while roaming, the WLAN architecture defines the ability for the WLAN device to establish a secure tunnel to the PDG. For example, users on an unsecured wireless connection (such as Starbucks Wi-Fi) are able to talk on the phone using the cellular network simultaneously while browsing the Internet over the Wi-Fi network. On a secured network, the WLAN device will establish an IPSec tunnel utilizing Internet Key Exchange version 2 (IKEv2). IKEv2 permits the same EAP method used for authenticating WLAN access to be repurposed for tunnel authentication.

11.2 GAN

A Generic Access Network (GAN) is the standardized version of Unlicensed Mobile Access (UMA) specified by 3G/LTE. GAN architecture leverages WLAN architecture; however, rather than using the IPSec tunnel to transport native IP packets from the WLAN device, a GAN presents a complete access system. The new system allows cellular control messages to be sent over TCP as well as circuit-switched media to be sent over the Real-Time Transport Protocol (RTP) rather than time-division multiplexing (TDM). Figure 11.3 shows the functionality of legacy network architecture and the flow of messages that were sent over TCP.

Legacy network signaling was transported over TCP/IP connections, using Generic Access Resource Control (GA-RC), Generic Access

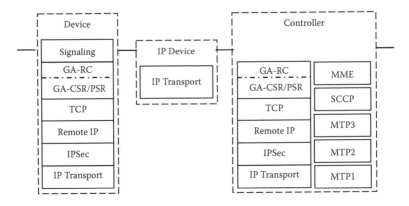

FIGURE 11.3   Generic access signaling transport.

Switch Resource Control (GA-CSR), and Generic Access Packet-Switched Resource Control (GA-PSR) protocols. GA-RC is a protocol that allows the GAN to communicate with a controller and supply an application keep-alive in an effort to enable the controller to estimate when a mobile user has transitioned outside of a WLAN coverage area. The protocols GA-CSR and GA-PSR effectively set up connections that transport packets on the circuit between the device and controller. These protocols also sustain both communication and transmission between the GAN and cellular nodes.

Figure 11.4 shows the same use of the legacy IPSec tunnel, but allows circuit-switched media to be sent over RTP.

Although GAN architectures were initially designed for Global System for Mobile Communications (GSM) and General Packet Radio Service (GPRS), circuit-switched networks have been optimized with each proto-col generation, supporting full-duplex voice telephony over WLANs. The

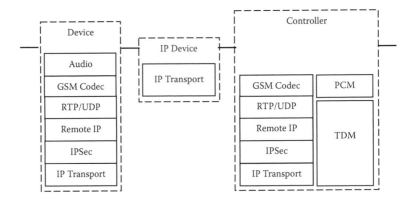

FIGURE 11.4   Generic access voice transport.

legacy infrastructures are only able to continue operating if the mobile device is able to simultaneously communicate over WLAN and cellular networks, by establishing an IPSec tunnel to register with the controller and accept assignment of a temporary identity. Assuming the cellular network is on the same broadcast domain, native mobile devices have the capacity to recover neighbor information; however, they will be unable to decode GSM intelligence and report information over the network. Consider traffic in the reverse direction—the mobile device will report measurements for the neighboring cellular network since legacy base station controllers (BSC) to GAN controller communication was established over the WLAN.

Given traditional IPSec implementations, a relative amount of session counts can be established with large throughputs. Consequently, GAN implementations that were previously discussed can only operate voice up to 3G, as hundreds of thousands of similar IPSec sessions are required and require a much more massive scale to drive throughput. So, as IPSec transport traverses WLAN network address translation (NAT) functionality, the interfaces in Figure 11.4 expand the bandwidth requirements even further.

11.3 WiMAX

The past few mobile system standards were centered on LAN functionality. The Worldwide Interoperability for Microwave Access (WiMAX) delivers wide area network mobile device access for broadband IP services. If you are unfamiliar with WiMAX, consider your Wi-Fi coverage at home or in the office. The two operate on the same general principles, sending data from one device to another via radio signals. Wi-Fi's range is measured in hundreds of feet, and WiMAX has a coverage radius of approximately 30 miles for fixed stations and up to 10 miles for mobile stations. The WiMAX architecture defines the use of IEEE 802.16-2005, currently known as 802.16e, as it provides physical and network layer support. The first license for commercial use was in the third quarter of 2006, whereas 802.11 technologies are for use for unlicensed radio bands. Mobile WiMAX air interfaces adopted orthogonal frequency-division multiple access (OFDMA) for streamlined multipath performance in deployments lacking line of sight.

WiMAX architecture is similar to the cellular networks discussed in this book with strategically located base stations using point-to-multipoint architecture to deliver services over a multiple-mile radius. When considering WiMAX, site surveys are performed, and based on the survey

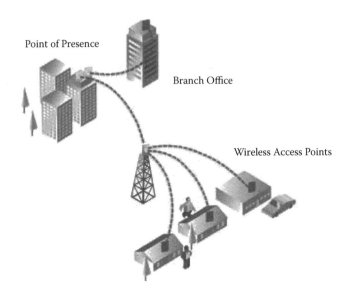

FIGURE 11.5 WiMAX is intended for wireless metropolitan area networks (MANs) and can provide service over greater distances than Wi-Fi.

results, an engineer can create a design based on equipment and architecture. Figure 11.5 illustrates the service capabilities WiMAX can offer for both residential and corporate deployments.

A WiMAX tower station can connect directly to the Internet using a high-bandwidth, wired connection. It can also connect to another WiMAX tower using a line-of-sight microwave link. Figure 11.5 illustrates a branch office connecting over a backhaul link to the point of presence, which obtains its access directly from the tower.

There are two main components for a WiMAX system:

- WiMAX tower: Consider the concept of a cell tower; a WiMAX tower can provide coverage to a large geographical area and serve both residential and business deployments.

- WiMAX receiver: Similar to a home or business having a router or switch to accept the service, a receiver or antenna serves as the component accepting the service for WiMAX, as illustrated in Figure 11.6.

Speaking from experience, deployments with high population density will generally limit capacity rather than range due to the available

FIGURE 11.6   Pictured from left to right, base station transceivers facilitate communication between networks and devices. Subscriber units provide connectivity to WiMAX networks.

spectrum. Base stations are backhauled to the core network by means of fiber or point-to-point microwave links to available fiber nodes, or via leased lines from similar carriers. When working for the stock markets, a backup deployment was conducted to test WiMAX capabilities over long distances. Since this was done just as WiMAX was released, we were not certain whether rain or snow would affect the signal or speed, let alone if matching engines would operate efficiently enough to trade. The only precipitation that affects a WiMAX deployment is heavy rainfall, and lightning poses a completely separate situation. When working with controlled indoor environments, deployments are more straightforward. For those who enjoy challenges, consider horizontally polarized signals must endure a higher degree of attenuation than vertically polarized signals as they propagate through the wireless channel under the influence of heavy rainfall. Subsequently, higher frequencies are more severely affected by cross-polarization and signal attenuation. System availability is reduced and the environment is rendered unstable due to the radio link performance degradation, resulting in an estimated 10% reduction in coverage. Throw a lightning strike into the equation, and your network is down until a technician can climb up and replace the damaged hardware. Thus, this type of link is only recommended as a backup link or if the geographic area does not offer other high-speed services.

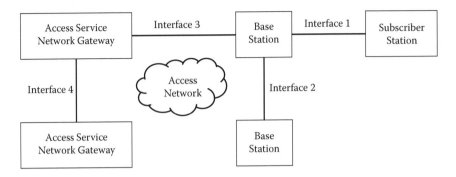

FIGURE 11.7   WiMAX system architecture.

Figure 11.7 displays the WiMAX system architecture with a subscriber station (SS), base station (BS) and Access Service Network Gateway (ASNGW).

The operation of the interfaces is as follows:

- SS and BS correspond to IEEE 802.16-2005 specifications over interface 1.

- Inter-BS communications are deployed to support seamless transport over interface 2.

- Interface 3 between ASNGW and BS provides control plane tunnel management, including the capacity to establish, modify, and release Generic Routing Encapsulation (GRE) tunnels.

- ASNGW mobility support is shown over interface 4.

Unlike cellular systems, WiMAX is defined as pools of resources between two completely separate organizations. IEEE 802.16 defines signaling between the subscriber station (SS) and the base station (BS); WiMAX defines the signaling between the BS and the ASNGW across interface 3, shown in Figure 11.8. Unlike cellular systems, where non-access stratum messages are defined, which allows wireless terminals to exchange signaling messages with core network components, no direct signaling is defined in WiMAX.

11.3.1 WiMAX Physical Layer

Since WiMAX uses an orthogonal frequency-division multiplexing (OFDM) transmission-based system, which contrasts with other cellular

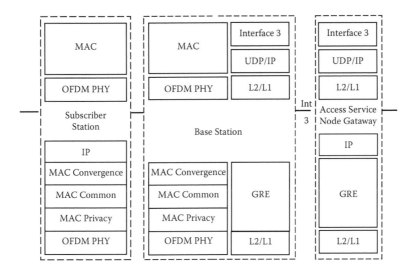

FIGURE 11.8   Generic access voice transport.

systems described in this chapter, the transmission scheme is scalable and can be used with a variety of channel bandwidths. One advantage of OFDM is allowing different users to transmit over varying portions of the broadband spectrum, or traffic channel. OFDM enables the use of high-speed data and video transmissions and other multimedia communications used by a variety of other broadband systems, such as DSL and Wi-Fi. Table 11.1 specifies the physical layer data rates at varying channel bandwidths. Operational parameters, such as data rate performance, vary depending on the defined parameters. Channel bandwidth and the modulating coding schemes used are the two parameters to focus on for WiMAX, but also consider the number of subchannels, oversampling rate, and OFDM guard time.

WiMAX also supports a coding and modulation scheme, allowing the scheme to change on a burst-by-burst basis per link, depending on channel conditions. A mobile device can provide the base station feedback on the downlink channel quality based off the channel quality feedback indicator. The base station will estimate channel quality based on the received signal quality, and make determinations for the uplink. Table 11.2 lists the various coding and modulation schemes supported by WiMAX.

TABLE 11.1   Physical Layer of WiMAX Data Rates at Various Channel Bandwidths

Channel bandwidth	1.25 MHz	3.5 MHz	5 MHz	10 MHz
Physical mode	128 OFDM	256 OFDM	512 OFDM	1024 OFDM

TABLE 11.2   WiMAX Coding and Modulation Schemes

	Uplink	Downlink
Coding	**Mandatory:** Convolutional codes at rates 1/2, 2/3, 3/4, 5/6 **Optional:** Convolutional turbo codes at rates 1/2, 2/3, 3/4, 5/6; repetition codes at rates 1/2, 1/3, 1/6, LPDC	**Mandatory:** Convolutional codes at rates 1/2, 2/3, 3/4, 5/6 **Optional:** Convolutional turbo codes at rates 1/2, 2/3, 3/4, 5/6; repetition codes at rates 1/2, 1/3, 1/6, LPDC, RS codes for OFDM-Physical
Modulation	BPSK, QPSK, 16 QAM; 64 QAM optional	BPSK, QPSK, 16 QAM, 64 QAM; BPSK optional for OFDM-Physical

Modulation and Code Rate	1.5 MHz		3.5 MHz		5 MHz		10 MHz	
	DL	UL	DL	UL	DL	UL	DL	UL
BPSK, 1/2	0.946	0.326	Not applicable					
QPSK, 1/2	0.504	0.154	1882	0.653	2520	653	5040	1344
QPSK, 3/2	0.756	0.230	2882	0.979	3780	979	7560	2016
16 QAM, 1/2	1008	0.307	3763	1306	5040	1306	10,080	2688
16 QAM, 3/4	1512	0.461	5645	1958	7560	1958	15,120	4032
64 QAM, 1/2	1512	0.461	5645	1958	7560	1958	15,120	4032
64 QAM, 2/3	2016	0.614	7526	2611	10,080	2611	20,160	5376
64 QAM, 3/4	2268	0.691	8467	2938	11,340	2938	22,680	6048
64 QAM, 5/6	2520	0.768	9408	3264	12,600	3264	25,200	6720

Physical Layer Data Rate (Mbps)

FIGURE 11.9   Physical layer modulation and coding schemes associated with WiMAX.

As mentioned above, data rate performance varies based on the operating parameters. Figure 11.9 illustrates the physical layer modulation and code rates operating at either the uplinks (ULs) or downlinks (DLs).

Having an OFDMA-based physical layer, both up- and downlink resources can be shared between users; discrete users can be allocated subsets of OFDM subcarriers in both the time and spatial domains (when optional advanced antenna capabilities are selected and used). These subsets, or subchannels, represent the minimum frequency resource that can be allocated to a user.

11.4 LTE

An alternative to WiMAX was under way in an attempt to optimize the already deployed IP-based architecture, and to limit the threat posed to cellular vendors and operators. Long Term Evolution (LTE) and Evolved UMTS Terrestrial Radio Access Network (EUTRAN) were the new IP-based access systems proposed after WiMAX. The specifications for EUTRAN were published in 2009 by 3GPP Release 8.

Requirements listed by 3GPP 25.913[*] for EUTRAN include the following:

- Ability to be deployed with flexible spectrum allocations from 1.25 to 20 MHz

- Average user throughput being three to four times the rates achievable with HSDPA previously specified in 3GPP Release 6

- Reduced EUTRAN user plane latency of less than 5 milliseconds in unloaded conditions

- Ability to support a large number of always-on users, up to 200 users for 5 MHz spectrum allocations

- Reduced state transition time between power-saving mode and active states to less than 100 milliseconds

- Instantaneous peak data rates of up to 100 Mbps in the downlink and 50 Mbps in the uplink (assuming a 20 MHz channel bandwidth)

The requirements listed above led to the definition of a new EUTRAN architecture together with a new, flatter, all-IP network, termed the Evolved Packet Core (EPC).

11.4.1 EUTRAN

The EUTRAN architecture aligns with Evolved High-Speed Packet Access (HSPA+) and was based on 3G networks with higher speeds for the end user, comparable to LTE networks. Both EUTRAN and HSPA+ steer radio-specific user plane functionality into an Enhanced Node B (ENB), including Packet Data Convergence Protocol (PDCP), Radio Link Control (RLC), MAC, and PHY protocols, as shown in Figure 11.10.

---

[*] 3GPP 25.913, "Requirements for Evolved UTRA (E-UTRA) and Evolved UTRAN (E-UTRAN)."

User Equipment		Enhanced Node B		Int 1
IP				
PDCP		PDCP	GTP	
RLC		RLC	UDP	
MAC		MAC	IP	
EUTRAN PHY		EUTRAN PHY	L2	
			L1	

FIGURE 11.10   EUTRAN user plane protocol, distributing functions into the ENB, so soft handover is not supported.

ENBs connect to the EPC using the Int 1 interface, with control plane functionality communicated across the mobility management entity (MME) interface, which transmits from the tower to an edge router. With this architecture, shown in Figure 11.11, the EUTRAN includes multi-homing, whereby a single ENB can be adjacent to multiple MMEs and serving gateways (SGWs). The interface transporting the MME traffic is based on IP transport using the Stream Control Transmission Protocol (SCTP), referenced in RFC 2960, which provides guaranteed delivery of application layer messages.

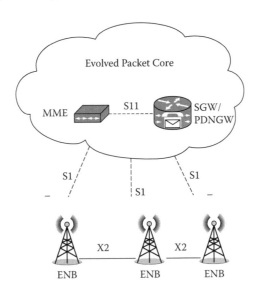

FIGURE 11.11   EUTRAN architecture.

Many network engineers employ continuity checks between the EPC and ENB. Even though SCTP heartbeat mechanisms could be used for the same purpose, this technique could only be used over the same interface to which our adjacent equipment is connected. Thus, we implement bidirectional forwarding detection* (BFD) over multihop paths in an effort to provide a low-level continuity check that can operate over any type of IP transport network.

Referring back to Figure 11.11, ENBs are interconnected using the X2 interface. The X2 interface is used to manage handovers between ENBs. On a hard handover, the X2 interface allows transient tunneling of packets between base stations, so it can handle packets delivered to an old base station once a user has moved to a new one.

11.4.2 Securing EUTRAN

The S1 and X2 interfaces are vulnerable to attack if an untrusted IP transport network is used for radio access network (RAN) backhaul. Therefore, engineers define IPSec on both X2 interfaces between ENBs, as well as the S1 interface, between the ENB and Security Gateway at the perimeter of an untrusted network. This proposed segmentation requires the use of multiple IP addresses on the ENB, allowing for separation of X2, S1, and management traffic.

11.4.3 EUTRAN Physical Layer

The LTE physical layer is similar to WiMAX, where it has embraced a multicarrier approach using OFDMA in its downlink. However, LTE uses single-carrier frequency-division multiple access (SC-FDMA) in its uplink. When comparing OFDMA, SC-FDMA vastly improved the peak-to-average power ratio (PAPR) attributes. Consider an engineer who must determine the linearity requirements of the mobile's power amplifier. The selection of SC-FDMA and its respective lower PAPR will allow LTE users at the edge of the cell to operate with a high PA efficiency. See Table 11.3 for additional attributes of the LTE physical layer.

The LTE physical layer is also capable of operating at either short or long cyclic prefixes. In this example, LTE would support multicast/broadcast over a single-frequency network. In such network deployments, multiple cells transmit a time-synchronized waveform, permitting the terminal to pool emissions from multiple cells. So, the longer cyclic prefix allows larger differential delays between cells involved in multicell transmission.

---

* RFC 5883, "Bidirectional Forwarding Detection (BFD) for Multihop Paths."

TABLE 11.3   Physical Layer Attributes of LTE

Advanced Antenna Systems	MIMO 2 × 2, 4 × 4
Base station synchronization	Frequency synchronization (FDD and TDD), time synchronization (TDD and MBSFN)
Channel bandwidth	1.25–20 MHz
Forward error correction	1/3 rate convolutional and turbo coding
Modulation type	QPSK, 16 QAM, 64 QAM (optional in UL)
Multiple access technique	Downlink: OFMDA
	Uplink: SCFDMA
Subcarrier spacing	15 kHz
TDMA frame duration	10 ms with 1 ms subframe

11.4.4 Load Balancing

Network redundancy is supported in LTE, referred to as S1-flex, where one EUTRAN system acts as the parent system to multiple Evolved Packet Core (EPC) networks. Depending upon the configuration, loads can be shared across network links and devices if they belong to the same EPC operator, or corroborate sharing if the EPC devices belong to autonomous network operators. As a device attaches to the EUTRAN, the device and service provider routes it to the respective core network device.

11.5 BACKHAUL

With the push for all-IP services, mobile Internet and enterprise mobile deployment demands continue to rise. More bandwidth is required, and current infrastructures are not optimized to handle these types of traffic. Legacy infrastructures were designed to handle voice and a trivial amount of data. Today, there is a need for network technologies that can efficiently handle voice and data traffic, yet are cost-effective.

IT managers and directors have found themselves in a competitive market where mobile operator's voice component of average revenue per use (ARPU) continues to decline. Consequently, new mobile service products must be offered in an attempt to slow down customer turnover where the grass is greener on the other side. Operation costs continue to escalate faster than revenue figures, and 25% of those operational costs come directly from infrastructure.[*] While a converged voice and data infrastructure using IP-based backhaul can deliver new multimedia at affordable costs, the engineer must scale for growth, or the network performance and availability will be crap within 1 to 2 years of its deployment.

---

[*] Infonetics Research, "Mobile & Wireless," 2014.

11.5.1 Network Design

Consider the following key elements when designing backhaul architecture:

- Simplify the network and provide a single technology that runs over a variety of transports, including legacy media, native Ethernet, or Ethernet over microwave links. Use IP Ethernet networks for service convergence, which provides the ability to deliver next-generation services over a single infrastructure.

- Mobile subscriber traffic is dominated by data, and will continue on that trend. Therefore, legacy TDM infrastructure is suboptimal. IP radio access networks (RANs) provide statistical multiplexing, making efficient use of the bandwidth available.

- Utilize IP over Ethernet to allow for easy infrastructure upgrades between 10 Mbps, 10 Gbps, and 100 Gbps.

- Higher-bandwidth services are becoming less expensive than the T1, DS3 legacy circuits. Not only are they easier to deploy, but they also provide differentiation opportunities for an engineer to be creative and segregate traffic.

Reference Figure 11.12 when designing all-IP mobile backhaul architecture, as described above. Metro Ethernet networks are widespread in commercial areas, providing carrier class and high-bandwidth networks. Be sure to price the services out, and just because a carrier is not in your building doesn't mean you should rule it out. A customer/tenant has the ability to bring any service into a building with a proper design plan. You will be surprised how many carriers have been scratching at the doors to provide a presence in new territory.

In preparation for the future converged networks, standards have already defined IP or Ethernet interfaces for most access technologies. To be successful, interfaces should be defined considering most of the current radio technologies, such as TDMA, GSM, CDMA, CDMA2000/EV-DO, and W-CDMA/UMTS.

Some networks have already migrated over to Ethernet infrastructures. Base stations are aggregated and transported over a metro Ethernet network, connecting to the central controller. Several carriers have already rolled out the network design just described, and are being used to provide services to business and residential customers over the same network.

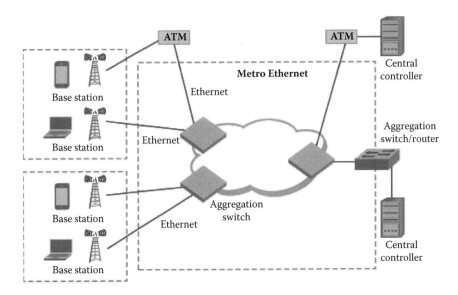

FIGURE 11.12   Mobile infrastructure deployment utilizing higher-bandwidth carrier services to scale for growth.

Such networks also provide layer 2 and layer 3 virtual private network (VPN) services to customers. So, other locations utilize microwave links as an effective, high-speed transport solution.

11.6 SUMMARY

This chapter identifies several cellular wireless systems, covers deployment approaches for each system, and describes the benefit of delivering an all-IP access system. There is still a need to deliver legacy traffic on top of IP traffic, so tunneling the legacy traffic over the GAN is generally the best practice. The two mobile broadband systems, WiMAX and EUTRAN, share downlink characteristics as well as offer scalable channelization approaches to support the operation of varying channel bandwidths. Since there are true all-IP systems with defined IP interfaces between base stations, it is imperative to properly scale and route traffic in order to deliver increasing traffic needs, from Mbps to Gbps. There are privately held access systems capable of delivering traffic in excess of 100 Gbps. For the reader of this book, the majority of infrastructures hover around 1 Gbps, and although there are obviously cost differences with optical networking, the functionality and design are comparative.

Traffic Delivery

Lauren Collins

CONTENTS

THE SUCCESS OF AN APPLICATION can make or break the amount of users who download, use, and recommend an app. Similarly, the performance, reliability, and availability of mobile network architecture and its traffic flow are just as important, if not more important, to the success of an app. Chapter 2 described use cases for location services and also introduced the level of functionality behind the framework for location services. Obtaining user location from a mobile device is oftentimes complicated. Not only are there multiple platforms to consider when developing, but there are also multiple options to choose from when building on top of a platform application programming interface (API). In this chapter we will cover one way to build these capabilities into your app using the classes of the `android.location` package.* The model described in this chapter, pictured in Figure 12.1, provides multiple entry points for a single app and invokes the individual use of components. Chapter 3 listed the various operating systems (OSs), discrete platforms to build apps on particular devices, and touched on the fundamental concepts to consider when choosing a platform, framework, or device. For the developers using framework API, or for those who are just starting to add

---

* "Location Strategies," Android Developers, n.p., 2012, http://developer.android.com/guide/topics/location/strategies.html (accessed May 26, 2014).

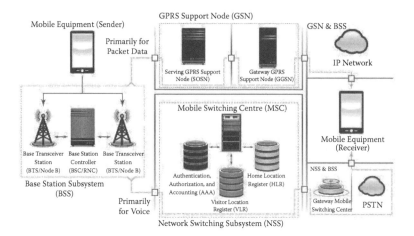

FIGURE 12.1 Mobility switching architecture, supporting signaling exchanges on behalf of the attached mobile device. When traffic is segregated by type (e.g., data and voice), a distributed architecture maximizes efficiency and allows for expansion.

location awareness to their apps, you should strongly consider using the Google® Location Services™ API, and this chapter will cover Google Maps and Google Location Services API.

The left side of Figure 12.1, base station subsystem, was described in detail throughout Chapter 11. The next section will focus on the remaining components of the illustrated mobile architecture.

12.1 GPRS SUPPORT NODE AND NETWORK SWITCHING SUBSYSTEM

In a Global System for Mobile Communications (GSM) architecture, a mobile station will communicate with a base station system (BSS) through the radio interface. The BSS connects to a network and switching subsystem (NSS) by communicating with a mobile switching center, pictured in the lower right corner of Figure 12.1.

Mobile stations are comprised of two components: the mobile equipment (ME) and subscriber identity module (SIM). A mobile station can also use a component referred to as terminal equipment, which can be a mobile device or a laptop connected to the ME.

SIMs can be

- Perforated card, containing an additional piece (plug-in SIM)
- Smart card
- Smaller variation of plug-in SIM

Each SIM is secured by a personal identity number (PIN) between four and eight digits. The PIN will be loaded initially by a network operator or at the time your subscription starts and will prompt the user to enter a PIN. The SIM contains relevant subscriber information, including a list of short dialing numbers, short messages received, and the names of preferred networks for service. If the number is consecutively entered incorrectly three times, the SIM is blocked and the MS cannot be used. In the case the SIM is blocked, a user must correctly input the eight-digit PIN unblocking key.

SIM data can be accessed and changed using software on a PC. During normal operation, subscriber-related data are sent to the ME; however, once a SIM is removed or a MS is deactivated, the data are deleted.

The NSS is comprised of switching functionality, mobility management, and subscriber profiles. There are basic switching capabilities within the NSS, but those are performed by the MSC. A MSC will communicate with neighboring networks outside of the GSM using a common signaling protocol. The approximate location of a MS is maintained by a visitor location register (VLR), which creates an interim record for the mobile device on the visiting system. Information for the visiting subscriber is stored in order for the MSC to provide voice and messaging service to the visiting device.

12.2 ANDROID LOCATION SERVICES

Knowing where the user is allows an application to be more efficient when delivering information to the user. When developing a location-aware application for the Android™ platform, you are able to utilize GPS and Android's network location provider to acquire the mobile user location. Although GPS is the most accurate, there are several downfalls: it does not always work outdoors, it quickly consumes battery power, and it does not return the location as fast as the user would like to see. Android's network location provider determines user location by cell tower and Wi-Fi signals, providing location information whether the user is indoors or outdoors, responds faster, and uses less battery power.

The following lists how to obtain user location on the Android platform:

- Get the user location by means of callback.

- Indicate to receive location updates from `LocationManager` (Location Manager) by calling `requestLocationUpdates()`, passing it a `LocationListener`.

- `LocationListener` must implement multiple callback methods the Location Manager calls when the user location changes or when the status of the service changes.

The following lists how to define a `LocationListener` and request location updates:

```
//Acquire a reference to the system Location Manager
LocationManager locationManager = (LocationManager)
this.getSystemService (Context.LOCATION_SERVICE);
//Define a listener that responds to location updates
LocationListener locationListener = new
LocationListener( ) {
  public void onLocationChanged(Location location) {
  //Called when a new location is found by the network
location provider.
  makeUseOfNewLocation(location);
  }
  public void onStatusChanged(String provider, int
status, Bundle extras) {}
  public void onProviderEnabled(String provider) {}
  public void onProviderDisabled(String provider) {}
};
//Register the listener with the Location Manager to
receive location updates
locationManager.requestLocationUpdates (LocationManager.
NETWORK_PROVIDER, 0,0, locationListener);
```

The first parameter in `requestLocationUpdates()` is the type of location provider you use (in this case, the network location provider for cell tower- and Wi-Fi-based location). The frequency can be controlled where your listener receives updates with the second and third parameters. The second is the minimum time interval between notifications, and the third is the minimum change in distance between notifications; setting both parameters to zero requests location notifications as often as possible. The last parameter is `LocationListener`, which receives callbacks for location updates.

To request location updates from the GPS provider, substitute `GPS_PROVIDER` for `NETWORK_PROVIDER`. Requesting location updates from both the GPS and the network location provider can be done by calling `requestLocationUpdates()` twice, once for `NETWORK_PROVIDER` and once for `GPS_PROVIDER`.

12.2.1 Request User Permissions

In order to receive location updates from `NETWORK_PROVIDER` or `GPS_PROVIDER`, request user permission by declaring either the `ACCESS_COARSE_LOCATION` or `ACCESS_FINE_LOCATION` permission, respectively, in your Android manifest file:

```
<manifest … >
  <uses-permission android:name=android.permission.
ACCESS_FINE_LOCATION" />
  …
</manifest>
```

Note: If using both `NETWORK_PROVIDER` and `GPS_PROVIDER`, `ACCESS_FINE_LOCATION` includes permission for both types of providers. (Permission for `ACCESS_COARSE_LOCATION` includes permission only for `NETWORK_PROVIDER`.)

12.2.2 Challenges in Determining User Location

There are multiple explanations why location readings are inaccurate or contain errors. Some include the following:

- User movement: You must account for constant user location updates by reestimating user location every *n* often.

- Variety of location sources: GPS, Wi-Fi, and cellular can all provide details pertaining to a user location. Determining which one to use and rely on is a game of balancing speed, efficiency, accuracy, and battery life.

- Accuracy: A location that was obtained 15 seconds ago from one source may be more accurate than the newest location from another source. Realize and accept that location sources are not consistent in their accuracy, so best practice is to estimate locations from multiple location sources.

12.3 SUMMARY

Understanding the functionality of a mobile network system is important when configuring or troubleshooting location services. This chapter defines the mechanism and systems in which location services use, and also provide, information to assist you in facing challenges in obtaining a reliable location reading. Additionally, we provided concepts that can be used in your application to supply the user with an accurate and responsive geolocation experience.

CHAPTER 13

Proximity Communications

Lauren Collins

CONTENTS

THE GOAL OF PROXIMITY COMMUNICATIONS is to deliver specific content, offers, promotions, and a more captivating shopping experience based on the precise location of a mobile device. Innovative, agile application developers are embracing the proximity communication space, yet they have not perfected a digitally engaging shopping experience. Chapter 2 introduced the concept of a proximity-based shopping experience, where a mobile device could be tracked from a prior shopping purchase or by viewing items online. Once the network recognizes the mobile device's MAC address, push advertisements can be sent based off proximity to the store and even a product shelf.

Once this concept is perfected, the potential of proximity marketing is fascinating for all involved: retailers, manufacturers, and shoppers. The task at hand is daunting, as improper timing, communication, or targeting of consumers can be detrimental to the success of both the retailer and the manufacturer. All of us have been exposed to inaccurate targeting or delivery of a message at an inappropriate time; however, you may have actually seen this work properly.

In order to refine the experience and implement proximity communications properly, here are some components to consider:

- Technology: What percentage of shoppers do not have the latest mobile devices, or the most recent OS version to support this technology? What if the shopper does not have Wi-Fi or Bluetooth turned on? How can we gauge whether a shopper is passing by a store vs. browsing?

- Timing: What timing and frequency of messages will strike the proper balance between enticing and irritating? If sensors are deployed across a large store, a shopper could be inundated with messages if he or she spends an hour at one store but browses multiple sections of the store.

- Content: What message format will shoppers want to see (or avoid) while shopping: audio, video, etc.? What types of content will interest a shopper: reviews, coupons, prior purchases, online shopping carts?

- Performance: The average shopper will only wait up to 5 seconds before abandoning a site and going elsewhere.[*] Adaptive web design sites load, on average, in 2.6 seconds, while responsive web design sites took over 4.3 seconds to load, and sometimes take up to 6 seconds. The inconsistency in load times can be attributed to the responsive web design sites being over 1.2 MB in size, while the adaptive web design sites, on average, were 791 KB.

As technologists, the mobile industry not only integrates fascinating technology, but also prompts the analysis of thousands of inputs applied to hundreds of complex rules. This presents a challenge in the successful delivery of coherent and effective information at accurate intervals. Proximity communications not only applies to the use of mobile devices while shopping, but also leads us to a question: As an industry, how do we implement solutions leveraging technology, expertise, and relevance?

13.1 RADIO FREQUENCY IDENTIFICATION (RFID)

Radio frequency identification (RFID) is a wireless tracking system that utilizes radio frequency electromagnetic fields to transfer data. There are several purposes for RFID, initially for identifying and tracking tags attached to a particular item or object. RFID is similar to bar

---

[*] Chantal Tode, "How Amazon, PayPal Gain Competitive Advantage on Mobile via Adaptive Design," Mobile Marketer, July 8, 2014, http://www.mobilemarketer.com/cms/news/software-technology/18163.html (accessed July 15, 2014).

code identification; however, the main differences are that RFID does not employ line-of-sight reading that bar coding depends on, the tag may be embedded in the tracked object, and RFID performs over greater distances than bar code scanning. High-frequency RFID systems (850 to 950 MHz and 2.4 GHz) offer transmission ranges of more than 90 feet, though wavelengths in the 2.4 GHz range are absorbed by water (e.g., the human body) and therefore have limitations.

13.1.1 Tags and Readers

The concept and design of RFID tags and readers is quite intriguing. Two-way radio transmitter receivers send a signal to a tag and read its response. The RFID tags may be active, passive, or battery-assisted passive (BAP), which is only activated in the presence of a RFID reader. Passive tags have the potential to emit more radiation than the other RFID types due to the power level for operation tripling that of active or BAP RFID tags. Tags come in many shapes and sizes and are programmable. While programmable tags have multiple uses and configurations, read-only and read/write tags allow for very specific data to be written to the tag. Tag information is stored in nonvolatile memory and is composed of two parts: an antenna to receive and transmit the signal and an integrated circuit to store and process information. RFID readers transmit an encoded radio signal to the tag, and the tag responds with identification information or other relevant information, such as a serial number or stock number. Figure 13.1 shows a forklift with one reader and two antennas. The antenna in the lower circle, located above the forks, locates a RFID tag embedded in the warehouse floor to identify a row. The antenna located at the top left reads the RFID tag placed on the racking system upright to identify the rack location. And the top right circle shows the reader, which allows real-time communication to the screen with the operator, located in the seating compartment.

RFID is used in some of the following applications:

- Track inventory, animals, and people
- ID badge to enter workplace
- Toll road and contactless payment (e.g., Mobil Speedpass, McDonald's, American Express ExpressPay)
- Airline baggage and boarding pass tracking
- Sensor networking

FIGURE 13.1 Forklift-mounted RFID antennas and reader allow for real-time communication between the operator and inventory location in manufacturing plants.

RFID tags are in practically every device that has the ability to capture, track, and record personal information, along with the movement of data pertaining to a trail, or history, of a person's activities. Parking garages now have cameras that take pictures of license plates and store the vehicle information in a database for years to come. Toll transponders are also used as tracking mechanisms in cities that utilize RFID to monitor traffic patterns. And if you do not own a transponder or use the parking garages, there is another tracking device attached to your car. Tire manufacturers insert RFID tags that can be read from around 20 feet away from a reader.

13.2 NEAR-FIELD COMMUNICATION (NFC)

Near-field communication (NFC) is a short-range wireless RFID technology that makes use of interactive electromagnetic radio fields, rather than direct radio transmissions. NFC supports short-range communications over wireless and is utilized for pairing of devices, sharing of information, and banking or shopping transactions. Mobile devices possess microscopic chips, shown in Figure 13.2, and are not in every device just yet. In 2014, only 800 million people are using NFC technology on mobile devices for banking purposes, but in 2019 over 1.75 billion mobile users are projected to engage in this service.*

Banking is not the only industry using NFC technology. The healthcare industry aims to verify the integrity of patient visits as well as reduce the number of fraudulent claims by implementing NFC tags and NFC mobile devices to provide an "indisputable and auditable" record of visits made to patients.† This technology will have the ability to prove that a caregiver was physically present in a particular place for a required length of time. Marc Bielmann, HID Global's vice president of identification technologies, states, "HID Trusted Tags satisfy Electronic Visitor Verification (EVV) requirements by leveraging the near ubiquitous nature of NFC-enabled mobile devices to enable a highly secure EVV system that is both easy to use and impossible to defraud."

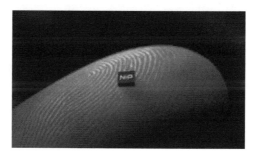

FIGURE 13.2 A near-field communication (NFC) chip by NXP is pictured on a fingertip.

---

* Rian Boden, "1.75bn to Use Mobile Banking—NFC World+," Juniper Research, July 11, 2014, http://www.nfcworld.com/2014/07/11/330268/1-75bn-use-mobile-banking (accessed July 15, 2014).

† Sarah Clark, "HID Global Aims to Cut Medicaid Fraud via NFC—NFC World+," n.p., July 8, 2014, http://www.nfcworld.com/2014/07/08/330222/hid-global-aims-cut-medicaid-fraud-via-nfc (accessed July 15, 2014).

13.3 BLUETOOTH

Bluetooth is a short-range wireless technology utilizing UHF radio waves in the industrial, scientific, and medical (ISM) band from 2.4 to 2.485 GHz.[*] The Bluetooth standard, IEEE 802.15.1, is no longer maintained but is managed by a special interest group (SIG). In order to implement the technology, there are strict qualifications that must be met along with a number of patents. And once the technology is approved for implementation, the license is only good for that qualifying device. Currently, Bluetooth version 4.1 is the latest specification announced by SIG. The update includes new features improving consumer usability with the support for LTE, developer enhancements, and bulk data exchange rates.[†]

There are a number of concerns associated with Bluetooth, such as security vulnerabilities. Encryption is not mandatory for Bluetooth versions prior to 2.1, and several regular operations actually require encryption to be turned off for properly functionality. If anyone uses Bluetooth on their mobile device to play music while driving in the car, it is a good idea to have a password for pairing. There is usually a default password, but changing that default password is strongly recommended.

13.4 BODY AREA NETWORK (BAN)

Body area network was initially defined by the wireless standard IEEE 802.15, a communication standard optimized for ultra-low-power devices and operation on, in, or around the human body (but not limited to humans) to serve a variety of applications, including medical, consumer electronics, and personal entertainment. In 2012, IEEE 802.15.6 was optimized to address and compensate for the effects of a body on network performance. This standard was implemented to enable a new generation of implantable devices, and assist in the development of new opportunities for delivering better healthcare, as well as support other innovative uses for wearable computing devices.[‡] Data rates up to 10 Mbps exist to support healthcare services, as well as the components matching the set of extensive and innovative personal entertainment.

---

[*] "Welcome to Bluetooth Technology 101," Fast Facts|Bluetooth Technology Website, n.d., http://www.bluetooth.com/Pages/Fast-Facts.aspx (accessed July 12, 2014).

[†] "Updated Bluetooth 4.1 Extends the Foundation of Bluetooth Technology for the Internet of Things," Press Releases|Bluetooth Technology Website, n.d., http://www.bluetooth.com/Pages/Press-Release-Detail.aspx?ItemID=197 (accessed July 15, 2014).

[‡] Shuang Yu, News & Events: Press Releases, IEEE-SA, n.p., May 21, 2012, http://standards.ieee.org/news/2014/iot_workshop.html (accessed July 15, 2014).

BANs represent many other technologies and uses associated with the body sensor network (BSN). BSN technology represents the lower bound of power and bandwidth from BAN use cases. Wireless personal area network (WPAN) is another group that has a need for the standards under IEEE 802.15, utilizing devices inside and in close proximity to the human body. The task group associated with IEEE 802.15 drafted a (private) standard encompassing a large range of possible devices. Consequently, the task force is providing mobile device and application developers with guidelines to operate under, balancing power and data rate.

13.5 SUMMARY

Technology is an extension of humanity and is a part of our everyday lives. Therefore, anyone aged 2 to 90 is learning to work smarter, not harder. Every day technology emerges, and our younger generations will continue to develop and improve upon existing engineering. Computer literacy is not going to take you very far these days, but being the best will. Knowing how to leverage the communication standards will lead to tremendous personal growth, opportunity, and even quality of life.

IV

Security and Data Analysis

Mobile Encryption

Lauren Collins

CONTENTS

PROTECTING BUSINESS AND HOME COMPUTERS from viruses and other threats has become a necessity. Best practice is to install virus software prior to connecting the cable and connecting to the Internet. So why are mobile devices overlooked? We store our passwords, perform online banking transactions, log personal medical data, and have our personal calendar on smartphones and tablets. Mobile encryption software is much like traditional security and encryption software installed on desktops and laptops at the office and in homes. Browsing websites can be done with ease without having to worry about opening any potentially harmful websites or downloading problematic apps. Mobile encryption software not only has the ability to encrypt files, but it can also filter text and phone messages, among many other features. Keep in mind that this chapter focuses primarily on Android™ mobile encryption software. Be sure to check the supported platform prior to purchasing a product and ensure your device is supported.

In order to select the proper mobile encryption software application, there are distinct areas of emphasis as well as the common useful features, such as phone locators and backup options. The first step in protecting your mobile device is to set a password, which protects against the most fundamental hacking and intrusion attempts. Mobile device encryption software protects against possible online threats through app downloads or web browsing. Typical software offers security and encryption for documents, password protection allowing for remote access to your device, and the quarantine of threats in select apps. The software runs in the background, providing real-time protection, and does not affect use of the device or usual operation.

Mobile encryption application features offered are as follows:

- Remote lock and wipe
- Phone locator
- Website filters
- Scan downloads
- Phone and text blocking
- SIM card removal protection
- Application locking

In addition to protecting your mobile device, performing regular backups and setting restore points is a great idea. It will assist tremendously in replacing data in the case of a lost, misplaced, or stolen device. If someone is able to surpass the first level of defense, your password, and the layers of encryption, at least you will be able to protect the information that was stored on the mobile device.

14.1 SOPHOS

One of the few mobile encryption software apps that are supported on the iPhone® and iPad® is Sophos™. At the time of publication, version 2.5 supports the configuration of data leak prevention (DLP) rules to securely access and export data from a particular set of content providers. Figure 14.1 shows some of the supported content providers; Dropbox and Google Drive are the most commonly known.

14.2 SYMANTEC

A popular player in the security market, Symantec™ requires a managed account on a Symantec Encryption Management Server with the Lightweight Directory Access Protocol (LDAP) Directory Synchronization feature enabled. LDAP is a vendor-neutral application protocol for accessing and maintaining distributed directory services. The Symantec PGP

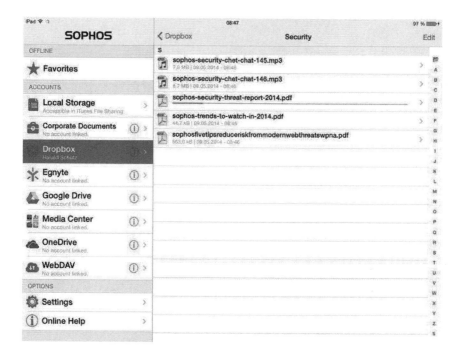

FIGURE 14.1   Sophos is a mobile encryption app for iOS and iPad that encrypts, stores, and uploads files within the app.

viewer, shown in Figure 14.2, decrypts messages and attachments, verifies digital signatures, and complements an enterprise mobile strategy.

Users are able to access and view encrypted email, attachments, and messages on the mobile device, ensuring the information remains protected until it is decrypted. Subsequently, regardless of whether the device is lost or stolen, the information remains protected and encrypted until it is decrypted.

14.3 KASPERSKY

The most attractive element to Kaspersky™ is its comprehensive range of protection, pictured in Figure 14.3. Most software is easy to use and does not require much, if any, assistance. However, in case you need to access customer support, Kaspersky has chat, email, phone support, and an expansive blog as options. The features of Kaspersky's remote phone locator set it apart from the competitors. After a stressful day at the office and a night full of barhopping, it is common to lose your mobile device. Using Kaspersky's web portal, the remote phone locator uses the mobile device's GPS and can either sound a siren or take a picture of the person

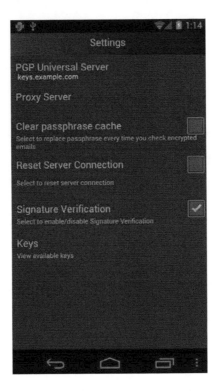

FIGURE 14.2 The Symantec mobile encryption app settings menu allows the user to select the features for protection.

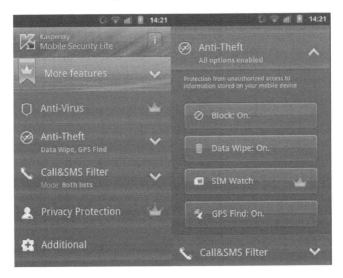

FIGURE 14.3 Kaspersky Premium mobile device encryption software offers the most comprehensive protection in the market.

who has the device, referred to as Mugshot. And if the SIM card has been changed, the phone has the ability to notify you of the new phone number.

14.4 SUMMARY

Mobile encryption software was designed to keep the data safe on your mobile device. Even though it is imperative to password protect your mobile device as a first-level defense, mobile encryption protects the information accessed and stored if the device is lost or stolen. Digital signing, or trust verification, is important to protect devices from a man-in-the-middle attack.

CHAPTER 15

Mobile Forensics

Scott R. Ellis and Josiah Roloff

CONTENTS

15.1 INTRODUCTION

Chapter 9, "Mobile Standards," went into great depth and detail about the history of mobile, including that of its weaponization. It did not, however, make any mention of criminalization of mobile. We thought we would save that for the chapter where it belongs, which is here. Aside from criminal matters, there have been several very influential decisions regarding the relevance of cell phones in civil matters. As of 2014, cell phone forensics is no longer the "new kid" on the block that it was a decade ago. Ten years ago, the primary challenge in cell phone forensics was finding a connector that would allow you to connect to the cell phone that you had, which may be a model that you had neither heard of nor ever seen previously.

Unless you live under a rock or have not been reading previous chapters because you have simply stumbled across this book and opened to this page randomly, by now you understand the usefulness that mobile has in your life. You probably have also figured out that in order to provide all of these useful features, all of these touchstones in your life, these points where virtual and real worlds intersect, your mobile phone needs to receive information about the real world. In a landmark decision in 2007,[*] Magistrate Judge John Facciola ordered Cafe Asia to preserve images that had been stored on a mobile device:

> One important constraint is the admissibility of the discovery being sought. Defendant asserts that the images, if relevant, are discoverable under Rule 26 even if inadmissible at trial. This holds true, however, only if the images "[appear] reasonably calculated to lead to the discovery of admissible evidence." Fed.R.Civ.P. 26(b)(1)....
>
> To the extent that defendant aims to use the graphic content of the images to establish plaintiff's "own" standards of behavior, Mot. to Compel at 1 (emphasis in original), the images themselves are the "end game" of the discovery request. As such, the question of discoverability is inseparable from admissibility, and a determination is necessary of whether, under Federal Rules of Evidence 403 and 412(b)(2), the probative value of the images substantially outweighs their prejudice. This determination is best made by the trial judge either pre-trial or in limine at trial, and for that reason I will order that the images be preserved pending a ruling on

---

[*] *Smith v. Cafe Asia*, 246 F.R.D. 19 (D.D.C. 2007).

their admissibility by Judge Robertson. Moreover, because Judge Robertson is entitled to a robust and fully informed debate over the admissibility of the images—a debate that cannot occur where only one party has the benefit of having seen them—I will order plaintiff to permit one attorney, designated by defendant, to inspect the images.

The analysis differs where the discovery is sought by defendant to corroborate the testimony of its witnesses that plaintiff willingly shared the images. Defendant believes that the images, if flaunted by plaintiff, are probative of whether the taunts (however tasteless) were innocuous teasing, and whether the Yim e-mail (however lewd) was playfully welcome. Plaintiff, however, denies having willingly shared the images with his co-workers. The specific content of the images may speak to the credibility of defendant's witnesses' testimony as to the nature of the images, and as to the nature of discourse between plaintiff and his coworkers.

Often, this information gets stored locally. And whether it's an inappropriate picture taken by someone of someone else at work that demonstrates harassment, or if it is evidence that you were, in fact, texting on your mobile when you got in that wreck, the data that land on a mobile phone have direct relevance and bearing on the decisions that a judge will make. It is for these reasons and more (more people have mobile devices, storage size increases, and increased portability of data) that evidence from mobile devices has increased so much in recent years.

Often, when a forensicologist has to testify in a hearing, there may be several other hearings before his. The judge's criminal docket is rarely lacking in people to fill it. In 2006, while waiting to testify in a similar case, I saw the power of a cell phone in the courtroom. A woman was trying to get a restraining order, and she was complaining about a message that her ex-boyfriend had left for her. A misunderstanding between the judge and the woman ensued. The judge did not understand the technology. I was actually about to ask my attorney to intervene when the woman finally got it through the judge's head that she had the cell phone with her, and she could play the message for him. *He was about to dismiss the case due to a lack of evidence.* He didn't understand that voicemails could be retrieved and played at will. She played the message, and not only did she get her restraining order, but the judge issued a warrant for his arrest (it was a rather nasty message, and it involved some sort of threat).

It is still not uncommon for judges and attorneys to display an abhorrent aversion to all things technical. This is why they hire experts. I have worked as one of the experts for many years, and while I seldom practice forensics today, I maintain my certification and believe that I can still lend some clarity to the topics. It is with that in mind that I have taken an existing guide to mobile forensics and carved out a thread that is a simplified and clarified instruction set. That is what this chapter is: it is one forensicator's guide to the National Institute of Standards and Technology (NIST) Mobile Forensics, Special Publication 800-101, and has my thoughts and impressions intertwined within it.

The relationship between the Department of Homeland Security and NIST is unclear; however, one thing is certain: this chapter must begin with a disclaimer. Just because something is published in this chapter is not a guarantee or a requirement that it be adhered to by any government organization. Whether it be chain of custody or procedural issues, going toe-to-toe with a federal investigator in a courtroom is not an effective or winning strategy.

Ultimately, because so many of today's judges and prosecutors are so (perhaps intentionally so) technophobic, in this author's experience they tend to err on the side of prejudice and side with the prosecution in any procedural disagreement. This is perhaps the most challenging aspect of professional testimony. Some judges very much want to hear the testimony from experts on both sides, and will ask questions and are genuinely interested and fascinated. Some other judges, however, are very technically adverse, and have no foundation to grasp even the simplest of concepts. I once testified as a neutral expert in a federal court in Alabama, and watched as the opposing expert (my report had come out in favor of the plaintiff, so the defense brought in an "expert") was slowly dismantled by the judge. As he talked about his qualifications, the opposing expert fumbled a bit. He had no knowledge of back-end database architecture, but he was able to suggest that his experience some 20 years ago writing BASIC should qualify him. The judge asked him what BASIC was, and the man explained that it was a programming language. The judge asked why it was called BASIC, and my opposition explained, "It's an acronym, but I don't recall for what." They then dived into the testimony. Later, after I had been sworn in, the judge went through the process of qualifying me. This was simply a formality, as he had already previously qualified me to testify as a neutral witness. After discussing my qualifications and qualifying me, he fired his first question at me: "I bet you know what BASIC stands for, don't you?" I did. And therein lay the victory in the case. From that moment on, the judge sided with my report 100%, and as I walked him through it, it became very clear that he understood and was paying attention to everything that I was saying. I had located hundreds of thousands of dollars worth of "orphan" database transactions in the defendant's database. These were detailed sales records, and there was no doubt that these were actual sales because the schema of the database proved it. As the hearing closed, the judge informed the defendant and the plaintiff to retire to a conference room and work out a settlement. If he had to make a judgment, he explained, the defendant would be most unhappy with the result. Mobile systems are no different than any other computer system. They contain databases of information about things like financial transactions, they contain "smoking gun" documents, and they require specialized knowledge to access them.

The best place for the forensic expert is in civil litigation, where he is on equal footing with the competition. Federal criminal court is a place where both the prosecutors and the judges have a nearly fanatical belief in the infallibility of the system, and cannot be bothered with facts—even when the facts clearly demonstrate that evidence should be tossed out, and that it had been altered due to carelessness.

Mobile forensics is a practice that is replete with opportunities for very serious mistakes—from accidental wipes of cell phones to confusion of evidence, mislabeling, tainted evidence, and lost/misplaced/misfiled evidence. Furthermore, the criminals that are leveraging the cell phone networks are not stupid.

This chapter dovetails my real-world practical and hands-on experience doing forensics and testifying in court to create a rubric against the backdrop of the NIST guide. Throughout this chapter, text from the guide is intentionally reprinted, and is followed by commentary or explanation. To differentiate between my thoughts and impressions and those of NIST, I have included the text from the publication as extracts and footnoted the page that it came from.

To begin with, the guide suggests that the agency is ambivalent about the technology that it uses, and that it implies no endorsement of one particular cell phone manufacturer over another.

> Certain commercial entities, equipment, or materials may be identified in this document in order to describe an experimental procedure or concept adequately. Such identification is not intended to

imply recommendation or endorsement by the National Institute of Standards and Technology, nor is it intended to imply that the entities, materials, or equipment are necessarily the best available for the purpose.[*]

Additional contributors include the following:

- Ronald van der Knijff and his colleagues at the Netherlands Forensic Institute and,

- Svein Willassen at the Norwegian University of Science and Technology for their assistance on technical issues that arose in our work.

- Rick Mislan from Purdue University.

- Chris Sanft from the SEARCH Group, and all others who assisted with our review process.[†]

The exact nature of these contributions is not known, but the following is true about each contributor:

- Ronald van der Knijff—Contributing author, *Handbook of Cybercrime Investigations*, first edition in 2002, and the last known edition, 2007.

- Svein Willassen—In 2005, published a paper on internal memory analysis of cell phones. He also developed a tool called SIMCon, which recovered deleted SMS messages, and was later acquired by SIMCon, which was later acquired by Paraben Inc. and included in Paraben Device Seizure.[‡]

- Rick Mislan—From Purdue University, now a visiting professor with Rochester Institute of Technology (RIT). Since 2000, he has consulted with dozens of police departments and federal agencies.

- Chris Sanft—From the SEARCH Group, and a trainer with Access Data and since 2008.

---

[*] Wayne Jansen and Rick Ayers, National Institute of Standards and Technology (NIST) Mobile Forensics, Special Publication 800-101, p. 3.

[†] Ibid., p. 4.

[‡] http://www.willassen.no/.

Each of these people provided a great deal of expertise, and their knowledge and experience enriches this document and lends it the sort of peer-reviewed expertise that such a document would possibly require to be, in and of itself, admitted as evidence and treated with the same level of respect (and hopefully more so) as the dissenting and contradictory opposing opinion that you may face. Do recognize, however, that at times even experts can have differences of opinions. While the NIST guide has been cited numerous times in this chapter, there are portions of it that are in direct contradiction with this chapter's advice. You should always research for yourself any testimony that you plan to deliver, and while citing is good, it is also necessary that you, as the forensic examiner, conduct tests and verify facts for yourself, as things may change as the industry and the technology develop into the future.

15.2 OVERVIEW—GETTING STARTED WITH PROCESS AND PROCEDURE

Mobile phone forensics is the practice of preserving the digital evidence that exists on a mobile phone that is the subject of an investigation in a court of law. Mobile phones are, relative to the advent of wireless communications, a relatively recent invention. Because of the way they work, and due to the incredible and even more recent convergence of mobile with full-featured computing capabilities, traditional computer forensic techniques may be inadequate, irrelevant, or completely infeasible to execute against a mobile phone. For this reason, a single, authoritative publication is necessary that can adequately explain the technology and how it should be treated in a forensic manner.

Procedures and the tools that support them must be explained in depth. Forensicologists will encounter unknown technologies and situations that require proper handling. The NIST guide is not an all-inclusive guide of everything that you can run into, and nor is this abbreviated reference. This chapter provides the foundation of knowledge necessary to establish proper protocols and procedures, and of course, at *no time* should this chapter be misconstrued to be legal advice.

Every situation is unique, as are the experiences of the forensic specialists and the tools and facilities at their disposal. The judgment of the forensic specialists should be given deference in the implementation of the procedures suggested in this guide. Circumstances of individual cases; international, federal, state, local laws and rules; and organization-specific policies may also require actions other than those described in this guide. As always, close and continuing consultation with legal counsel is advised.

The following guidelines, then, become readily apparent. Organizations should develop both comprehensive policies and procedures. The difference between the two is subtle, but I have broken the NIST guidelines regarding this into two separate sections to clarify the differences. Procedures are checklists of things that people follow and do; policies are written documents that define when procedures should be executed, who should be doing them, and when they can be made:

- Step-by-step procedures should detail the following:
  - All routine activities with respect to preservation, acquisition, examination, and analysis.
  - Templates and procedures for reporting of digital evidence found on cell phones and associated media.
  - If it is a repeatable action, and is carried out in a forensic manner, it should be documented.
  - Maintain the chain of custody.
  - Store evidence appropriately.
  - Establish and maintain the integrity of forensic tools and equipment.
  - Be able to demonstrate the integrity of any electronic logs, records, and case files.
  - Develop a procedure that describes who conducts periodic reviews, and lists the documentation that needs to be reviewed. Provide also for the "sign-off." Someone needs to be ultimately responsible to ensure that this work was completed satisfactorily.
  - Explain and document the methods for examining every brand of mobile that comes into your shop.
  - Document the procedure to be followed when new models are encountered.
  - Include checklists of things that may indicate a failure—for example, how long does the collection of data from a certain phone of a certain size take? How much of a delta is acceptable?

If the USB port on the phone is damaged, or the normal connector doesn't seem to fit, what should be done?

- Comprehensive policy should be established that defines the following:

  - Establish a policy and assign the execution of it to a human body. Periodic review *and* whenever needed.

  - Define how you will support, identify, acquire, and train on mobile forensics toolsets. The surface area of mobile changes rapidly, and with the emergence of new applications that can be at the center of any lawsuit, how will you rapidly acquire the knowledge you need about the inner workings of an app?

  - Policies should be established regarding when procedures should be followed verbatim, or when working from memory is OK. When I used to work on C-130 aircraft as a USAF reservist, we had technical procedures that we would follow, step-by-step. It wouldn't do to miss a step when installing an electric fuel pump solenoid. Mistakes make planes crash, and when dealing with justice, mistakes can put innocent men in jail, free the guilty, and cause wrongful judgments on the orders of millions of dollars to be paid out in civil litigation.

  - Sometimes, mobile phones will arrive damaged. The bad guy may have thrown it out the window of a moving car, or perhaps an angry mistress stomped on it. Explain the degree to which repairs are allowed, and when they should be attempted. Some shops will specialize in this and have this capability; others may need to send it out.

  - Policy should demand that all policy be reviewed by a legal expert, and where policy is accompanied by a procedural set, a senior and qualified forensic expert should review it. The policies should comply with local and federal laws.

  - Policy and job descriptions should be documented, communicated, signed, and stored for easy retrieval. All policy should be reviewed by all staff, and all procedures should be tracked in a skills matrix that ensures that those who are executing the procedures have been fully trained and understand what is taking place on the phone and on the equipment, throughout the process.

- Unusual situations call for creative thinking, at times in stressful situations. New procedures that need to be executed in live situations should be collaborative, and the ability to respond quickly and make solid decisions should be tested and practiced.

When you find yourself in a court of law, you should be prepared to not only present a book of a listing of your procedures, but also discuss them in depth, and defend the validity of each step of each procedure. You should be able to demonstrate the last time that you reviewed the procedure. If it can be argued, it will be. When these guidelines don't exist, what will support the admissibility of evidence into legal proceedings? And don't worry if any of the above items seemed intimidating. I will review each one of them in detail in this chapter, and provide some sample frameworks for documents as well.

15.2.1 Attribution

The NIST guide was prepared for use by federal agencies. According to it:

> It may be used by non-governmental organizations on a voluntary basis and is not subject to copyright, though attribution is desired.
>
> Nothing in this guide should be taken to contradict standards and guidelines made mandatory and binding on Federal agencies by the Secretary of Commerce under statutory authority, nor should these guidelines be interpreted as altering or superseding the existing authorities of the Secretary of Commerce, Director of the OMB, or any other Federal official.

The purpose of this chapter is to interpret that guide, add value to it, and explain its processes and procedures in as near to layperson's terms as possible. I use shorter sentences, cut irrelevant content, add additional information, and ultimately produce a guide that is quick and interesting and can be easily referenced. I have removed excess words and rhetoric, and substantially trimmed down the guide.

This guide will not provide any step-by-step processes for any mobile phone models, and it will steer clear of mentioning any brands of mobile phone or software. Readers of this software are assumed to (possibly) be forensic examiners that are interested in learning mobile forensics. In that vein, then, it sets forth a framework that can be used to better understand the entire process of mobile forensics.

When a cell phone is encountered during an investigation, many questions arise: What should be done about maintaining power? How should the phone be handled? How should valuable or potentially relevant data contained on the device be examined? The key to answering these questions is an understanding of the hardware and software characteristics of cell phones.

This chapter provides answers to these questions, and insights related to the capabilities of mobile phones and the networks they use. But why do mobile phones require specialized knowledge to examine? Aren't they all the same?

> The NIST Guide ... gives an overview of the hardware and software capabilities of cell phones and their associated cellular networks. The overview provides a summary of general characteristics and, where useful, focuses on key features. Developing an understanding of the components and organization of cell phones (e.g., memory organization and use) is a prerequisite to understanding the criticalities involved when dealing with them forensically. For example, cell phone memory that contains user data may be volatile (i.e., RAM) and require continuous power to maintain content, unlike a personal computer's hard disk. Similarly, features of cellular networks are an important aspect of cell phone forensics, since logs of usage and other data are maintained therein. Handheld device technologies and cellular networks are rapidly changing, with new technologies, products, and features being introduced regularly. Because of the fast pace with which cellular device technologies are evolving, this discussion captures a snapshot of the cell phone area at the present time.[*]

Chapter 10 of this book provides an overview of the cellular networks in both the United States and Europe. A skilled mobile forensics examiner will have familiarized himself or herself with the both Code-Division

---

[*] Lily Chen, Joshua Franklin, and Andrew Regenscheid, *Guidelines on Hardware—Rooted Security in Mobile Devices*, National Institute of Standards and Technology, October 2012.

Multiple Access 2000 (CDMA2000) and Global System for Mobile Communications (GSM) networks.

It is of key importance is to understand that the network that the phone uses is not necessary in order for an examiner to acquire the subject phone. The term *acquire* is used frequently in many forensic texts, including this one. It doesn't mean acquire in the traditional sense, which means to get a hold of some property or some such thing. Rather, in forensic science, it means to copy (acquire), from a digital apparatus of any kind, all of the data that are created by that device by the user. Often, this includes bringing along a lot of other information with it, such as the entire operating system, all application data, as well as seemingly unused sections of the electronic storage.

A *forensic* acquisition is a process that endeavors, by the best of available technological means, to copy all of this electronic information from the apparatus and to storage media, and it intends to do this without any alteration to the data itself. Usually, this means that the device must be "imaged" in a way that is described as being *forensically* sound. Certain extenuating circumstances may make such an acquisition impossible. These scenarios will be discussed in this chapter; it should not be taken as the final word. Certainly between this writing and the time of publication, the ever-shifting field of forensics and mobile technology will converge and diverge multiple times on any number of topics addressed herein.

Before we move on from the subject of forensics and acquisitions, let's take a moment to understand what the word *forensic* means. Quite literally, and singularly, it means "of or belonging to a public forum [debate]." Now, since judges get to decide what is and isn't evidence in their court (forum), and they are not reliant or even conveniently predisposed to take a look at the NIST definition, nor listen to a lecture (debate) from a forensic computer expert on the forensic validity of evidence, it is important that the experienced investigator not rest the entire validity of his case on the poor forensic capabilities of the opposing counsel's examiner.

If a judge feels it is evidence, that it is relevant, material, and substantially affects the merit of the case, the sworn testimony of the expert presenting the evidence may be more than enough to establish and meet the requirements for evidence.

15.2.2 Preserving Digital Evidence in the Age of E-Discovery

Society has awakened in these past few years to the realities of being immersed in the digital world. I suspect we inhale the ones and exhale the zeroes. With that, the harsh realities of how we conduct ourselves in this age of binary processing are beginning to take form in both new laws and new ways of doing business. In many actions, civil and criminal, digital documents are the new "smoking gun." Along with the new federal laws that have opened the floodgates of accessibility to digital media, the sanctions for mishandling such evidence become a fact of law and a major concern. Citicorp recently paid out millions of dollars in fines because it said that most of the emails that were relevant to a case had been destroyed in the Twin Towers. It's like the grown-up version of saying "My dog ate it." Only, there's no dog and nothing devoured.

At some point, most of us (any of us) could become involved in litigation. "We live in a litigious society," and one out of every three people will be involved in litigation at some point in their lives. Divorce, damage suits, patent infringement, intellectual property theft, employee theft, fraud, and employee misconduct are just some examples of the kinds of cases we see. There are of course criminal actions, and criminal actions also allow for a great deal of digital dilemmas, but we won't get into that here. I've tried my best to leave my criminal career behind me. When it comes to digital evidence, most people just aren't sure of their responsibilities. They don't know how to handle the requests for, and subsequent handling of, the massive amounts of data that can be crucial to every case. And just like the proverbial smoking gun, digital evidence must be handled properly.

Recently, a friend forwarded me an article published about a case ruling in which a routine email exhibit was found inadmissible due to authenticity and hearsay issues. What we should take away from that ruling is that electronically stored information (ESI), just like any other evidence, must clear standard evidentiary hurdles. Whenever ESI is offered as evidence, the following evidence rules must be considered. A piece of paper that claims to be a printout of an email may very well be just that, but it could also be a Word document formatted to look just like an email. It's not hard, and anyone with a modicum of experience in word processing or who knows how to configure an email client can generate fake emails by the ream. What it comes down to is both authenticity and traceability. One must be able to retrace the steps with a witness at each node of transition.

In most courts, there are four types of evidence. Computer files that are extracted from a subject machine and presented in court typically fall into one or more of these types:

1. *Documentary evidence* is paper or digital evidence that contains human language. It must meet the authenticity requirements outlined below. It is also unique in that it may be disallowed if it contains hearsay. Emails fall into the category of documentary evidence.

2. *Real evidence* must be competent (authenticated), relevant, and material. For example, a computer that was involved in a matter would be considered real evidence provided that it hasn't been changed, altered, or accessed in a way that destroyed the evidence. The ability to use these items as evidence may be contingent on this and is why preservation of a computer or digital media must be done.

3. *Witness testimony*: With electronically stored information (ESI), the technician should be able to verify how she retrieved the evidence, and that the evidence is what it purports to be and should be able to speak to all aspects of computer use. The witness must both remember what he saw and be able to communicate it.

4. *Demonstrative evidence* uses things like PowerPoint, photographs, CAD drawings of crime scenes, etc., to demonstrate or reconstruct an event. A flowchart that details how a person goes to a website, enters her credit card number, and makes a purchase would be considered demonstrative.

For any of these items to be submitted in court, they each must, to varying degrees, pass the admissibility requirements of

1. Relevance

2. Materiality

3. Competence

For evidence to be *relevant*, it must make the event it is trying to prove either more or less probable. A forensic analyst may discover a certain web page on the subject's hard drive that shows the subject visited a website where flowers are sold and that he made a purchase. In addition to perhaps a credit card statement, this shows that it is more probable that the subject of an investigation visited the site on his computer at a certain time and location.

Materiality means that something not only proves the fact (it is relevant to the fact that it is trying to prove), but is also *material* to the issues in the case. The fact that the subject of the investigation purchased flowers on a website may not be material to the matter at hand.

Finally, *competency* is the area where the forensic side of things becomes most important. Assuming that the purchase of flowers from a website is material (perhaps it is a stalking case), how the evidence was obtained, and what happened to it after that, will be put under a microscope by both the judge and the party objecting to the evidence. The best evidence collection experts are trained professionals with extensive experience in their field. The best attorneys will understand this and will use experts when and where needed. Spoliation results from mishandled ESI, and spoiled data are generally inadmissible. It is the responsibility of everyone involved in a case—IT directors, business owners, and attorneys—to get it right. Computer forensics experts cannot undo damage that has been done, but if involved in the beginning, they can prevent it from happening.[*]

Electronically stored information (ESI), all of it, may contain useful information. When powered on, mobile phones will attempt, repeatedly, to contact and establish a connection with their master network because data about the phone need to be captured so that the subscriber service can properly bill and service the subscriber.

15.2.3 What Defines a Mobile Phone?

This is perhaps a ridiculous question to even ask in 2014, but as the boundary between mobile phones and mobile devices becomes murkier, the question deserves some attention. A mobile phone is any device that can be used to engage in mobile communications. For example, phone calls can be placed through a mobile device that is built in to my car. I can use my tablet for voice and video conferencing when I am on a Wi-Fi network. My landline has a box that is installed in my house that allows me to use it to make phone calls over the Internet from my mobile phone. The vectors of communications are increasing, and this means that the mobile forensic examiner has more considerations and work to do than ever. Today's mobile devices are every bit the mobile computing device, a

---

[*] Scott Ellis has provided expert consulting services for the defense in both state and federal jurisdictions.

micro-sized PC that possesses far more computing power than the computers of 10 years ago.

Some models of mobile phones and devices include additional slots for memory cards. Still others allow for this extensibility (albeit possibly limited) via connecting to the data/charging port. Very few (if any) models have separate power and data ports. They come in all sizes and shapes, from very little internal storage to many gigabytes, with enhanced security capabilities to no capabilities. In other words, there are so many offerings on the market that becoming an expert in the true sense of the word (someone who has nothing left to learn about a topic) is impossible. Most mobiles sport the following features, and regardless of how unsophisticated one may appear to be, the unfamiliar examiner should treat all devices as though they contain one or more of the following capabilities.

By 2014, most phones that are in the hands of all but a few sophisticated users will possess these capabilities:

1. Global positioning system (GPS)

2. Logging capabilities

3. Camera (video and stills)

4. Text message

5. Voice message

6. Email

7. Address book

8. Calendar

9. Web browsing

10. Portable database information—applications (apps) that may contain easily accessed data

11. Document review

These are really the most basic of features for the mobile units of today. While voicemail is a feature often included with most modern phones, often the voice messages are kept on a remote server and may be inaccessible without being connected. Some phones will download the message once it has been reviewed. So why then did I, just now, suggest that these great features are in the hands of all but the most *sophisticated* users?

There are three categories of people who opt for low-tech phones:

1. People who genuinely can't afford them.

2. Technophobes—these are people who either dislike technology of any sort or are paranoid that the government is watching them (and now we know that this was not really paranoia at all).*

3. Criminals and wrongdoers that are practicing antiforensic techniques. They use mobile disposable phones and cycle them out to prevent revealing information about their network, should they be apprehended.

Most of us, unless we have been living under a rock, are at least peripherally aware of the three to four major players in the mobile phone industry. These are the phones that are in the hands of the vast majority of professionals.

Mobile devices are rapidly expanding in breadth and scope. This chapter primarily concerns itself with traditional smartphone devices, but the informed forensicologist is aware of the data storage and communication capabilities that may be inherent in any of the following items:

- Wristbands

- Watches

- Tablets

- Mini-tablets

- Glasses

- Clothing (yes, some clothing comes embedded with arrays of useful electronics)

- Automobiles

- Digital cameras

- Game devices

---

* A web search for Edward Snowden, the former NSA contractor, reveals the depth and magnitude of U.S. government spying on its allies, enemies, and citizens. Recent allegations, unrelated to Snowden, suggest that the CIA actively spies on legislators; specifically, they raided the computer systems of Diane Feinstein's Senate Intelligence Committee.

The list is expanding. While it is true that the data storage capabilities of many of these items are nonexistent, it is also true that, 15 years ago, the data storage capabilities of mobile phones were also nearly nonexistent. Today, mobile phones and devices can store many gigabytes of data on both built-in flash and insertable cards. And, while automobiles don't turn up in the list of mobile "devices," they are mobile, and a frequent target of forensics. They present a very challenging field of forensic study.

This chapter will focus on mobile phones. In particular, it will focus on smartphones, which are phones that have one of about seven of the most common operating systems. There are many other types and models of regular phones that can also contain a wealth of information of investigational relevance, but the vast majority of phones, as well as the complexity, lies in smartphones.

One could certainly attempt to delineate the smartphone from its less feature-rich cousins, but to what ends? If a mobile device comes across your bench, you must examine it. Even more importantly (well, inarguably most importantly), you must not accidentally break or damage the phone. You must be able to identify, either at a glance or through microscopy, the connection type. Accidental forcing of the wrong connector on to a phone, especially if the connector is hot, can have serious consequences. OK, it won't kill anyone, but it could "fry" the phone and destroy any data on it that may, of course, have either exonerated or implicated the defendant.

15.2.4 Operating Systems

This chapter tries to remain brand agnostic, and discuss strategy instead. Despite that fact, there are a few mobile device operating systems. A good forensicologist will understand which type of phone pairs with which operating system, and will have a solid understanding of how to locate artifacts and put together a report for each of the systems. Think about this for a moment: you need to understand no less than seven operating systems, and how to examine them.

The thoughtful examiner understands his limitations, and seeks outside counsel as needed; if you choose to forge ahead; you should master one of these seven. Build an application for it. Dissect it. Subscribe to message boards and blogs about it. The other mobile platforms will become easier. Blackberry® and Symbian® have been placed at the bottom of this list because, short of some sort of revolutionary offering, Blackberry investigations may soon be a thing of the past and Symbian has been publicly deprecated. Because of my experience with iOS, you may notice that, throughout this chapter, the tips you see about iOS will be more prevalent than those for the other systems, with which the author has less intimate experience.

1. iOS (iPhone, iPad, and Retinal code paths)
2. Android (based on the Linux kernel—you know Linux, you got it made!) (wait—owned by Google!) (seen on Samsung Galaxy phones)
3. Windows Mobile (Microsoft)
4. WebOS (Linux based)
5. Symbian (Samsung, Motorola, Sony Ericsson, and Nokia)
6. Blackberry (Research in Motion (RiM))
7. Miscellaneous proprietary (non-smartphones)

15.2.5 Memory Considerations

Cell phones use several different types of random access memory (RAM) and NAND memory access. The architecture of how memory and information is stored will be covered in greater depth later, but only with respect to whether or not the information you seek will be available, whether it persists, and how volatile or susceptible to damage it may be. The actual architecture, while interesting, is hardly the place for a tactical forensic discussion and is better suited to academia or development of acquisition tools.

15.2.6 Subscriber Identity Module (SIM) Cards

There was a time when this author, when changing phones, just had to make certain that all of his contacts were stored on the SIM card, and not on the phone, when switching phones. I had two locations where I could store information. The SIM card is a way that cellular phone companies have of tracking their subscribers. It invoked portability, and allowed users to easily switch phones without long, bothersome reconfiguration sessions with a technician. What this means to the forensic examiner, then, is that having a cell phone without a SIM card is a less than perfect forensic situation within which to find oneself. Possession is 9/10 of the law (again, this is not legal advice), so the fact that the custodian had the phone in his or her possession is meaningful, regardless of SIM disposition. Modern cell phones cannot function without a SIM card. The immediate suspicion that a law enforcement officer should have about a missing SIM card is that the owner of the phone perhaps removed the card and threw it away, or possibly even ate it in an effort to hide or destroy evidence. That suspicion may be correct, or it may not be. Keep reading and I'll explain.

15.2.7 What's Stored on the SIM and What's Not?

This can be very different between phone models and carriers, and it is rapidly changing and evolving. Develop an understanding of how to capture and interpret this information. The Universal Integrated Circuit Card (UICC) carries a very small amount of information. The fact that they can contain relevant personal information is more a matter of mere annoyance than it is of any real consequence, but one that, nonetheless, should not be overlooked. It may contain information about last numbers called, and will contain information that may help correlate a phone to a user. For example, a user may suggest, "Hey, that's not my mobile, someone stole it and inserted their card into my phone." Well, the presence of contacts and shared, last dialed numbers, and other personal information about the individual being investigated may help repudiate such claims.

Caveat: Not *all* phones have SIM cards. It depends in part on the phone, and on the type of network, of which there are three, or maybe two and a half would be a better approximation.

15.2.8 A Brief Word about Mobile Networks (and SIM Cards)

TDMA, GSM, LTE, iDEN, Digital AMPS, CDMA2000, etc. Navigating the world of digital transmission technologies is pure alphabet soup. Chances are good, as a forensic examiner, you won't be quizzed on these things while on the stand, as they bear little relevance to the actual data on the computer, short of telling you how they might have arrived there, the protocol of that communication, and its security features. Primarily, there are two transmission technologies with which you should be familiar: Universal Mobile Technology System (UMTS) or GSM, and CDMA2000, which is a standard built off the earlier 2G CDMA work. GSM is used exclusively in most of Western Europe, while China and America have implementations of both. It's a bit of a Beta vs. VHS comparison situation. Perhaps Blu-Ray vs. HD DVD (high-density DVDs) is a more contemporary comparison.

More importantly, and most relevant to forensic examinations, CDMA2000 phones do not require a SIM card. GSM do. In the United States, AT&T and Sprint use GSM, and Verizon, T-Mobile, and U.S. Cellular use CDMA2000. Now, earlier I said "two and a half" because the next generation of CDMA2000, LTE, does require a SIM card. Another important distinction to make is that CDMA phones cannot make phone calls and receive data at the same time. What then does it mean if data clearly landed on a phone during an in-progress call? See this footnote for the answer.*

Interestingly, GSM is built on CDMA technology, but that's a topic for another chapter.†

Mobile phones, as they are called in much of the world, aka cellular phones in America, attach to networks that allow the phone to move between towers with a persistent connection. These networks contain the equipment needed to manage the mobile phones that enter and depart cell tower ranges. Chapter 9 of this book presents a guide to cellular infrastructure. To the forensicologist, the most important aspect of the network infrastructure is that cell towers have ideas, and if you are able to get the log files from the towers, they can help pinpoint a particular subscriber's location, and have been used often in the solving of crimes.

15.2.9 Software Tools

The tools that are available to conduct forensic examinations of mobile devices are largely geared toward the well-known operating systems. From the list above, item 7, proprietary systems, represents a discipline that is subject matter enough for a book on its own, whereas other material can be treated in a way that allows for the application of forensic disciplines with a unified approach.

The functionality of software tools designed for mobile devices is, in fact, very similar in utility to that of the tools used for stationary and portable computing (desktops and laptops). Proprietary phones, again, represent a unique challenge and have closed operating systems that can be difficult to understand. Nonforensic tools may be the only option available

---

* Many phones have a slot for a SIM card, even if it is not used. At airports in Europe, you will notice kiosks where you can purchase a SIM card and receive local rates, and download data and talk at the same time.
† Chapter 2, this book.

to communicate with some phones, and while they omit the typical hashes and double-checking of integrity functions that traditional tools offer, they are arguable in court when no other alternative exists. If the chain of custody is clear and concise, and the testimony of the expert is cogent, and the testimony is not based on data and information that were damaged or removed by the process, then the argument gains strength. For example, if a timestamp on a file is changed by a connection, one cannot argue or present evidence regarding the timestamp on that file.

15.2.10 NIST Tool Classification System

NIST classifies five levels of tools. The tools are classified by their difficulty in use, and the data that they are capable of recovering. Level 1 is the simplest, and level 5 represents the most complex and invasive of tactics. Figure 15.1 represents the triangle of mobile acquisition.

FIGURE 15.1   A standard four-number PIN lock screen.

- Level 1: Manual acquisition:

  - In this method, using a camera mounted above the cell phone, you navigate through every screen and photograph the evidence. Through careful photography, I once acquired several thousand screenshots of a computer system that defied examination. Later, because of the careful photography, I was able to use optical character recognition (OCR) on all of the photos and then compile the information from the screens into a spreadsheet that allowed for evaluation of thousands of orders. Yes, it took a long time, but when digital acquisition is impossible, manual methods are all that remain, short of writing an application program interface (API) where none exists.

  - Tools exist that can aid with this process and can assist in processing the information that is collected photographically.

- Level 2: Logical extraction:

  - The device is connected to a computer via wired or wireless technology, and commands are sent through the software interface to the device. Examiners should be cognizant of what will happen to the data when different connectivity types are used, as they can have different results. A forensic examination is rarely a place for experimentation. Judges should be notified when "dead ends" are hit. A good forensic examiner will inform counsel when his actions may destroy or alter data. This allows legal counsel to make informed decisions—they may ask you to proceed, may grant you the time to develop a solution, or may choose to preserve and pursue other avenues of investigation.

- Level 3 involves physically extracting and creating a copy of physical memory.

- Level 4: The chip is removed and is physically imaged:

  - This presents many challenges, as it's entirely possible that the chip will be encrypted and unreadable, or will have an undetectable file system. However, there is hope. Digital security systems present a similar dilemma at times. If the device is a popular one, there may be R&D staff in the United States that have, on hand, tools that they will give you that will allow you to connect and read the memory. This author once had a similar experience with a surveillance system file system. Going through the legal department is the best course of action, and you may not even need a subpoena.

- Another technique is to wire directly into the mobile device's Joint Test Action Group (JTAG) interface. You can then control the device's microprocessor directly to create an image.

- Level 5: "Micro-read," e.g., gate reading. This, according to the NIST guide, involves using a microscope to view the physical state of gates.

Level 5 is a premise put forth in the NIST paper. The notion that an electron microscope can be used, at any level of magnification, to "see" the physical state of gates is not one that, scientifically, can be entertained at this time. The guide goes on to say that this is the most invasive, sophisticated, and technical of approaches, and that nobody is known to be doing it. This author, with a university background in physics, finds it both improbable and unlikely that a microscope of any power can be used to evaluate the physical state of a NAND gate and see its position, which, in fact, is entirely electronic and not at all physical.

15.2.11 UICC (SIM Card) Tools

Being equipped with the capability to perform read-only acquisition of any and every type of memory card is critical. That the card readers be *one card only* and that technicians understand *card types* are equally critical. Given an array of five different types of memory cards, an experienced mobile device examiner will be able to easily identify each—as easily as you can identify the difference between a motorcycle and a car. Along with these tools should be a UICC personal computer/smart card (PC/SC) reader.

The following information may be available and recoverable from a UICC card:

- International mobile subscriber identity (IMSI)

- Integrated circuit card ID (ICCID)

- Abbreviated dialing numbers (ADNs)

- Last numbers dialed (LNDs)

- SMS messages

- Location information (LOCI)

- Deleted SMS messages

- Country and network operator codes

CSIM partitions on UICCs are becoming more prevalent for applications such as mobile money and securing healthcare data. The primary challenge of accessing these data lies in understanding and breaking their encryption. They will likely come of increasing importance in financial investigations.

There are a growing number of services that allow people to send and receive money via mobile, and also to complete point of sale transactions. These services require deep encryption, and it is at the level of the identity of the user—the SIM—where this occurs. By partitioning off portions of the UICC, individual financial information can be stored in a secure location that cannot be accessed by other apps. It may also be used to secure healthcare information.

15.3 PROTECTED DEVICES

Mobile devices, being mobile, are more susceptible to theft for many obvious reasons (because it is easy), and an unlocked phone can be a lucrative tool in the hands of a well-practiced criminal. To combat this, device manufacturers have implemented many security features. There are some techniques that can get around and recover data from devices that are protected. They fall into three general categories, which are covered in the following sections.

15.3.1 Software Attack

Some software tools provide a method for bypassing the security of many devices. The ability and success of these tools vary widely, and it should not be expected that all devices can be hacked in to. Many devices are extremely resistant, and include encryption capabilities as a standard offering. Sometimes, when a backup of a device exists on a PC that has also been collected, a procedure can be derived that will allow the examiner to back up the device and restore it to a separate phone without a password, and then that phone can be acquired.

15.3.2 Hardware Breaking

Some devices allow for the mobile device key to be read from a memory dump of certain devices, and this allows the phone to be unlocked and acquired. If a procedure cannot be located or developed easily, it may require some experimentation, which should be done on a test device. There are many specialized techniques, and they are almost always for a specific device, and will likely even be designed just for a particular device within a class. Do not anticipate or even expect that a particular technique will work on all firmware versions of a phone and all generations of it.

JTAG and flasher boxes may also be effective. These tools are covered below.

15.3.3 Social (Investigative) Hacking

There are some things that will likely always be true, at least until the capacity for humans to create their own passcodes has been removed.

Passwords, especially numeric ones entered into a number pad, will be easy to crack. Take your standard keypad interface: Which four number combinations do you think are most likely? Most likely, your subject's phone has one of the most common iterations of a keypad mnemonic and you can get it within a few tries. However, be mindful that a phone might be setup to self-destruct the data after too many wrong guesses. You should be aware of any apps that are "out there" that may allow a phone to be configured to self-destruct after just one wrong guess. You should be familiar with the security features of your phone.

Without even looking at a study or published list, let's take a guess at what the top 10 four-numbered codes are. These are not in any particular order; they are just the top 10 codes that I might personally choose if I were to not use a more secure code:

1. 1379
2. 1234
3. 2580
4. 2468
5. 3333

6. 9999
7. 1111
8. 7777
9. 5555
10. 0000

Beyond this, look to things such as the following as other common access codes:

1. Last four of social security number

2. Children's birthdays

3. Suspect's birthday

4. Address

5. Spouse's or girlfriend's or boyfriend's birthday

6. Last four of phone number—any phone number, especially a child-hood phone number

15.3.4 Smudge Attack

Some popular mobile phones have a graphically based password whereby the user swipes his finger in a preordained pattern of motions. I'm sure that you can imagine where this is going, right? Spraying the phone with certain chemicals, photographing it in the just-right light, or a as a last resort, boiling some methyl or ehtylcyanoacrylate (between 49 and 65°C [about 120 and 150°F]) in a box with it will cause those smudge patterns to light up like a Christmas tree in my backyard on the Fourth of July (my annual Christmas tree burn). Boiling cyanoacytate is also called the super-glue fuming method. As always, preservation is important. Doing this, and removing the smudge will reflect it, so be sure to treat this just as you would a fingerprint.

15.3.5 Flasher Boxes

Flasher boxes are essentially the moral equivalent to the memory dump capabilities of server and PC operating systems. They are a tool, developed by the original equipment manufacturer, for use in debugging. Much like a memory dump, they require specialized knowledge to operate and inter-pret. When a phone is damaged or compromised in some way, data is still present on the device, even if it's missing a SIM card. Make no mistake about it, though: when an automated tool is available for use, it should be used. Flasher and twister boxes are devices of last resort. They are an invasive tool, safe for use only by those who understand their use and are experienced in using them. Consult with legal counsel, and point them to the Department of Justice protocols for electronic crime scene investigation, which details the acceptability of invasive methods such as flasher and twister boxes.

There are some "gotchas" that must be well understood when using a flasher box:

- They are invasive.

- They are challenging to use.

- They do not allow for auditing or hashing of the evidence they collect.

- They may change data.

- There are dozens of types.

- Analyzing the data is very time-consuming.

- You may not be able to interpret the output data—it may be encrypted, and the skills and time required to reverse engineer the encryption algorithm may be out of the scope of this examination.

So when would you use one?

- When proper forensic avenues have been exhausted.

- When there are data that has been deleted that you know exist, and must retrieve in order to complete your examination.

- When the possibility to stop a crime from being committed exists.

- When you need to attempt to access data on the PIN card that is partitioned.

When nonstandard methods are used to conduct forensic activities, don't forget that *forensic* just means "suitable to be discussed and debated in a public way." The barrier for this would seem to be very low, and if a forensic examiner can provide reasonable evidence of how or why evidence may have been altered, this is usually enough (depending, of course, on what side she is on).

15.3.6 Joint Test Action Group (JTAG)

Formed in 1985, JTAG provides circuit board testing routines. JTAG is the acronym for what is known as the IEEE 1149.1 Standard Test Access Port and Boundary-Scan Architecture. JTAG is a standard that circuit board manufacturers use to allow for troubleshooting and connecting to their boards. It is a way whereby an examiner can connect to a device through test access ports (TAPs) and subsequently issue commands to the processor units mounted to the board. Some of these commands include instructions to "spill the beans," that is, to dump out all data stored on the device over the designated connection.

The process for this involves either using a known TAP or probing for ports and exploring the board, followed by soldering leads to the port. This is not for the light of heart—years of experience performing microsoldering are required, and the expert should also be very good at taking things apart without damaging them. "Jtagging" can be conducted against any device that has embedded flash memory, is supported, and has identifiable TAPs. Companies such as Binary Intelligence, located in Ohio, have the specialty skills on staff for this sort of work* (no affiliation).

15.4 STANDARD OPERATING PRECAUTIONS AND PROCEDURES WHEN CONDUCTING CELL PHONE INVESTIGATIONS

Rule number one is that there is no substitute for a proper Faraday cage. The bags that you see at trade shows can be expensive, and the larger they are, the more expensive they get. As a former (recovering) physicist, what I can tell you is that there is a basic principle behind how these cages work: a conductive layer

1. Reflects incoming fields

2. Absorbs incoming energy

3. Creates canceling fields

The bottom line is that any metal box works. Garbage cans, microwave ovens, old ammo cans, etc., can all be used for this, provided they are *conductive*. Not all "metal" boxes are made out of truly conductive

---

* See http://www.binaryintel.com/services/jtag-chip-off-forensics/jtag-forensics/.

material. For example, many stainless steel garbage cans are hardly electrically conductive at all and would not make suitable Faraday cages. If you are going to improvise your own Faraday cage, test it. Before building your Faraday cage, test the electrical resistance of the material with an ohmmeter. The resistance should read zero.

For example, I decided that I wanted to build my own Faraday cage using a paper bag and metal duct tape. The first thing I checked was the conductivity of the tape. Using the bag as my model, I then constructed my Faraday cage. As long as the bag remains closed, my cell phone is lifeless. If the metal tape at your local hardware store is not electrically conductive, you can always make a bag out of aluminum, and then cover it with metal tape to provide strength. For analysis in the field, consider investing in a lot of tinfoil and metal tape and (don't cringe) a car with an interior that you do not so much care about, complete with flaps for the windshields and windows. Either that, or you can buy a tent and, using metal tape, cover it with tinfoil. Your best bet is to fold the tinfoil like you fold a flag, into triple-ply triangles. Make enough of these triangular-shaped "tiles" to cover the surface area of the tent, and then apply using a strong epoxy adhesive. Test your adhesive on the fabric of the tent, first. When you are building this tent, just explain to your wife that you need a place to go where the aliens can't hear your thoughts.

Certainly, you could also identify underground parking lots or garages of some kind, or even the deep interior of a building that is dead, but this should be tested with each use. You never know when the proprietor may install a phone signal booster repeater. Never take it for granted that just because one cell phone does not receive signal in a location that you believe to be cellularly dead, other mobile devices will fail. Many forensic guides that are on the Internet suggest that you should use multiple devices to test your location. Either way, whether you build a 300 to 2000 MHz signal detector yourself à la Radio Shack style, or purchase one on the Internet ($300 to $500), you are better off with the peace of mind and *knowing* that your location truly is a dead zone.

15.4.1 Forensic Copies

More than once, I've received "forensic" copies of information from opposing counsel that were, in fact, forensic acquisitions that the opposing examiner had made of a nonforensically sound copy of something. For example, one investigator had attached to a network, copied all of the files from a network share to a local drive, and then tried to pass the work off as being forensic. It is far better to send the original copy, along with an explanation of what it is and how it came to be, that it is not forensic in nature. In this case, I imagine the investigator may have said something like:

> To whom it may concern. This hard drive contains a READ ONLY restoration of a file copy that was made under circumstances where a forensic acquisition was not possible. In order to preserve the information from any future issues, I have placed the files into an LEF, the password for which is ____.

Such a precaution can save you from having to answer difficult questions under oath. It's a disturbing notion, it's true, but being honest and accountable is always the best policy.

Forensic acquisition through means of a specialized tool will typically result in a "hash" being generated. A hash is typically a 128-bit hexadecimal character string that represents a digital "thumbprint" of a device. A simplified way of thinking about this would be to take this chapter and arrange all of its underlying binary into 128-bit-long strings. Each 128-bit string would then be XOR'd against the next 128-bit-long string, in the exact sequence of how the bits appear in the bit stream of the digital file that is this chapter. It follows, then, that due to certain mathematical and statistical certainties, the odds that another electronic file would match this *exactly* are truly astronomical. In reality, the most secure of hashes use a variable and nonlinear function with bit rotations and compression functions to ensure that the output file is, in fact, extremely likely to be unique. A number of good tutorials exist on the web that detail the exact process.[*]

Mobile devices, when on, are just like computers in that they will be constantly "doing stuff" to themselves that alters data. No two subsequent hashes will be the same. It follows then that a hash is more for the purposes of future data validation and retention than for verification of the device. If a subsequent hash is required, it must be understood that the phone will need to be powered up to collect the information and subsequent hashes will be different.

---

[*] A good example can be found at http://m.metamorphosite.com/one-way-hash-encryption-sha1-data-software.

15.5 WRAPPING IT UP

Finally, mobile device forensics is not for beginners. You should be thoroughly experienced in the art of forensics with standard office and consumer-grade equipment before attempting a foray into the world of mobile. *All* the procedures and protocols that apply to safeguarding information also apply to the field of mobile forensics. This field demands documenting, tracking, and paying meticulous attention to detail. Inability in this area is not a showstopper, though. Creativity and out-of-the-box thinking are in sad shortage in the forensic industry. In fact, one forensic examiner has been quoted as saying, "Forensics is the art of forgetting." What this really means is that the forensic examiner becomes somewhat locked in to a rather narrow spectrum of capabilities. No one forensic examiner can know everything, and so it is that she must specialize, and forget that she ever knew anything about other technologies, and she must become narrow and expert in one vertical of focus.

Mobile Device Security

C.J. Wiemer

CONTENTS

MOST OF US HAVE CONNECTED to a public wireless network at a coffee shop, an airport, and event at work. And if you have that network saved, your device will automatically reconnect next time without you even knowing. When connecting your mobile device to a wireless network, you want to make sure the network is secure. There are different levels of security that a wireless network can provide. Building on levels of security a network can provide, there are multiple types of encryption.

16.1 WEP

The oldest and least secure of these different security levels is Wired Equivalent Privacy (WEP) (introduced in 1999). This form of security for wireless networks is rarely used anymore due to its many security flaws. Some of these flaws include using a small 40-bit key that can be brute-forced, no key management, and no user authentication. Though you are authenticating to the network, this algorithm is MAC address based, and

MAC addresses can be sniffed, stolen, or easily spoofed. There are plenty of guides out there on how to crack WEP and gain access to a network that uses WEP. Do not connect to a network if it is using WEP.

16.2 WPA

The next level of security that came out was Wi-Fi Protected Access (WPA). This was introduced as an intermediary security protocol in anticipation of WPA2, which is the more secure and complex version of WPA and is what most networks use now because of this. WPA provides a more secure way of connecting to the network than WEP, getting rid of the unchanging encryption key that WEP was using and replacing it with the Temporal Key Integrity Protocol (TKIP). TKIP provides a 128-bit encryption key for each packet of the session, that is, each time a device tries to connect to the network, whereas WEP's encryption key does not change for the different sessions.

Two different modes exist, personal mode or preshared key (PSK) and enterprise or WPA-802.1X. PSK is most often used in small offices and homes. Since it is meant for a smaller amount of users, there is less complexity that goes with it, in that it doesn't require an authentication server, whereas enterprise does. A 256-bit encryption key is used by each network device when connecting to a WPA network using PSK mode. Enterprise mode means that the user is using a username and password, or digital certificate, to authenticate to a server (usually a RADIUS server) rather than using a preshared key.

16.3 WPA2

The 802.11i standard, or WPA2, was released in 2004 and improves on the security mechanisms that WPA uses. WPA2 has the same two different modes that WPA has, personal and enterprise. However, starting with WPA2, TKIP is no longer used. Instead, it mandates the use of the Advanced Encryption Standard (AES). Whereas TKIP was a stream cipher, AES is a block cipher. This is the major improvement with WPA2, as it is a more secure cipher.

16.4 HASHING

Hashing is a technique used to secure a user's password. It is essentially a complicated algorithm. The point of the hash functions is that they are hard to reverse, meaning if you put a string of characters through the algorithm to get a much more complicated string of characters, it would be hard to take that complicated string outcome and reverse it to find out what the original string was. These strings are normally passwords. The passwords are put through the hashing algorithm to result in a complicated string of characters to hide the original password.

Hashing is used as a means of authenticating a user's password. The server that the user is trying to authenticate to has a stored hash of all the user's passwords. When the user enters his or her credentials, the server will take the string of characters and run it through the hashing algorithm that it was set up to use. If the hash of the user's input exactly matches the hash the server has stored for the user, he or she is allowed successful authentication.

16.4.1 Case in Point

Bob's password is *Password*. The server will have the hash of *Password* stored in its database. For this example, the server is using the MD5 hashing algorithm, one commonly used with Android™ devices. For Bob, the server will have the hash dc647eb65e6711e155375218212b3964 stored in its database. Bob enters his password as *Password*, and the server runs that through the MD5 hash and compares its output with dc647eb65e6711e155375218212b3964. They match, so Bob is now able to access the server's data.

The point of using these hashing functions is to protect the user's credentials in case the server's database is compromised. If it is compromised, the hacker will only get the end result of the hashing functions, such as dc647eb65e6711e155375218212b3964. If the hashing algorithm is a secure one, it will be very difficult for the hacker to figure out what the password is based on that large string of seemingly random characters. However, each string of characters, when hashed, has a unique output. No matter how many times *Password* is typed using MD5, the output will always be dc647eb65e6711e155375218212b3964. To compare the results, here is what *password* would look like when using MD5 (difference is lowercase *p*): 5f4dcc3b5aa765d61d8327deb882cf99. The two strings look nothing alike even though the difference is only the case of the letter *p*. Having this knowledge, create your own hashes of commonly used passwords and compare them to hashes of the hacked database. For example, hash the password *Password*, compare it to the hash value for Bob's user, and notice the strings are identical. From this, we know that Bob's password is *Password*. Comparing these hash values is how hackers are able to derive user passwords.

There are many different hashing algorithms out there: SHA1, MD6, SHA256, etc. The difference between all of these is just the algorithm that the original string is put through to come up with the hash value. The more secure the algorithm, the longer it takes to output the hash value. Even though MD5 is technically broken, it is still widely used because it is somewhat secure and quick. The more that databases are compromised and their stored hash values are made available to the public, the more people are able to make comparisons of their own hash values to derive passwords. Thus, it is important for companies to use a secure hashing function, but equally important for users to choose complex passwords.

16.5 MS-CHAP

MS-CHAP is the Microsoft® version of the Challenge-Handshake Authentication Protocol (CHAP), and there are two versions of the protocol, MS-CHAPv1 and MS-CHAPv2. MS-CHAPv2 is the protocol that began being used in the Windows® operating system Vista. Do not confuse MS-CHAPv2 with WEP, WPA, or WPA2. CHAP does not provide a network level of security like the ones mentioned above. However, it is used within WPA and WPA2 to enable devices to communicate with each other. Before diving deeper into what MS-CHAP consists of, it is necessary to understand not only the underlying technologies and protocols that make it, but also those that preceded it.

For starters, it is a Point-to-Point Protocol (PPP), which is a data link protocol. This particular protocol is commonly used to establish a direct connection between two networking devices. Most Internet service providers (ISPs) use PPP for customer dial-up access to the Internet. It is able to provide connection authentication, transmission encryption privacy, and compression. Before a device is able to establish a point-to-point link, each device must send link control packets to configure the data link. Figure 16.1 illustrates the flow of communication of the data link layer. By default, authentication is not mandatory, but if it is desired, then the link must specify which protocol it wants to use. One such authentication protocol is CHAP.

CHAP is used periodically to verify the identity of the peer. It does this by using a three-way handshake that occurs with the initial link establishment. Once the link establishment phase completes, the authenticating authority sends a challenge message to the peer. The peer then responds with a value that is calculated with a one-way hash function. A hash value is a string of characters put through a specific algorithm, in this case RC4,

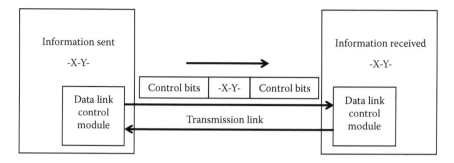

FIGURE 16.1 The data link layer ensures the communication on the physical link is reliable and efficient.

which results in a different string. Given this new string of characters, it is very difficult to reverse the algorithm to figure out the origin of the given string. Once the peer has calculated the hash value, it's sent back to the authenticator. The authenticator will then check this value against the value it came up with when it did its own hash calculation. If these values match, the authentication is acknowledged; otherwise, the connection will be terminated. To make this a continuous process, the authenticator will send a new challenge to the peer at random intervals and then have to repeat those same steps of authentication. This protocol authenticates a user or network host to an authenticating entity, similar to when you authenticate to your ISP.

Microsoft took this and extended the user authentication functionality provided on the Windows network to remote workstations. This became MS-CHAPv1, a Microsoft protocol to authenticate remote Windows workstations. This protocol provides the functionality to which local area network (LAN)-based users are accustomed to, while also integrating the encryption and hashing algorithm used on Windows networks. MS-CHAPv1 works similar to CHAP. First, the client requests a challenge from the server (the authenticating authority). The server abides the client's request by providing it with an 8-byte random challenge. The client uses the LAN manager hash of its password to derive three data encryption standard (DES) (a symmetric key encryption algorithm) keys. Each of these three keys is used to encrypt the challenge. All three of these now encrypted blocks are then concatenated into a 24-byte reply to the server. The client creates a second 24-byte reply using the Windows NT hash and the same procedure as before. The server, which contains the client's login credentials, uses the hashes of the client's password to decrypt the replies.

If the server's decrypted blocks match the original challenge that it sent the client, the authentication completes and sends a "success" packet back to the client.

Disadvantages and weaknesses for CHAP are

- Very weak and easy to crack.
- Hashed passwords are sent over the network, leaving them vulnerable to network sniffers.
- Does not limit the number of password retries allowed, making brute-force password guessing easier.
- One-way authentication (client is only one being authenticated).

MS-CHAPv2 solves the above disadvantages and weaknesses by providing two-way authentication, or mutual authentication. The protocol is more secure by doubling the number of bytes used for the password hash, from 8 bytes up to 16 bytes. Also, MS-CHAPv2 uses the SHA1 algorithm, which is harder to crack. MS-CHAPv2 is faster with the use of piggybacking, to decrease the number of packets used while authenticating.

MS-CHAPv2 no longer allows the weak LAN manager-encoded responses or password changes. As mentioned, MS-CHAPv1 only had one-way authentication, whereas MS-CHAPv2 provides mutual authentication between peers. It does so by piggybacking a peer challenge on the response packet, and an authenticator response on the success packet. An authentication scheme was introduced for the server that helps provide mutual authentication. This scheme is to prevent malicious servers from masquerading as legitimate servers. There have also been some changes to the different types of packets to fend off spoofing attacks. As shown in Figure 16.2, after the challenge has been sent, the client can now decide to continue with the communication based on the success or failure response given by the server. The client can also check the validity of the authenticator response, and may disconnect if the response has an incorrect value.

16.6 TYPES OF ATTACKS

16.6.1 Wireless Network Intrusion

Though choosing WPA2 as your wireless security level is secure, there are other vulnerabilities that can present themselves with setting up a wireless network, for instance, the use of Wi-Fi Protected Setup (WPS) on your wireless router. This is usually correlated with a button that is on the

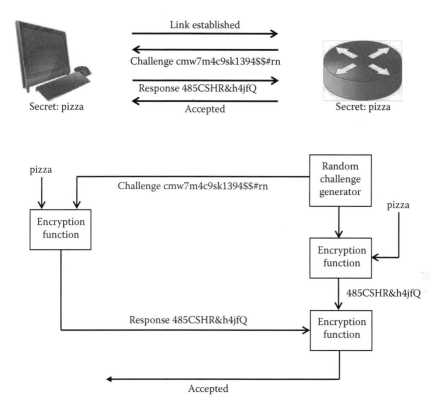

FIGURE 16.2 The working cycle of CHAP authentication. Once the link has been established, the authenticator sends a challenge message to the peer. Both the client and the server must know the challenge; in this case, "pizza."

router that when pressed, allows computers to connect to the router easily without any setup configuration. This feature is usually enabled by default on most popular routers currently out there.

How WPS works is by the use of a PIN number. This PIN number is hardcoded (cannot be changed) to each router. Usually to connect to the router, you would go on the device that you are trying to connect to the wireless network and enter in the PIN number that is associated with the router, usually on a sticker somewhere on the router. Since the PIN number is associated with the router and does not change, all the router is doing is confirming that the PIN number the wireless device is sending to the router matches that of the router. Once it confirms this, the wireless device now has access to the wireless network.

Since the PIN number is an eight-digit number, there are a limited number of combinations that it can be. In reality, there are only seven unknown digits because the last number of the PIN number is a checksum of all of the previous digits. Since there are seven unknown digits, this gives us 10,000,000 (10^7) possible combinations. The vulnerability that comes into play with this is that when the wireless device sends the PIN number to the router for it to confirm it is correct, the router will actually let the wireless device know whether the first half of the PIN is correct or not. With this being known, you have now cut down the number of guesses in half. This makes it an even more reasonable brute-force attack. All one has to do is to continue to send guesses of random PIN combinations to the router until it guesses the correct PIN number.

One can perform this brute-force attack to discover the router's PIN number by using a tool called Reaver, which is built in to the operating system BackTrack 5 R3 or Kali Linus, all free downloads. This tool will not only give you the PIN number of the router, but also the password to the wireless network, even if it's a WPA2 network. Also, remember that the PIN does not change on the router. Once you know the PIN, you will always be able to access the router. Even if the password for the wireless network changes, you can tell Reaver the PIN number it can start guessing at (given it's the already known PIN), and the program will spit out the new password to the network.

16.6.2 Network Manipulation

One of the most common and easiest attacks to perform is a man-in-the-middle (MitM) attack. This type of attack does not just apply to computers; it can be performed on mobile devices as well. Pictured in Figure 16.3, a MitM attack is when the attacker places its devices in between the victim's computer and the router. When the victim sends information to the router (checking email, logging in to an online account, etc.), the data are first passed through the attacker's computer, and then the attacker's computer passes the data along to the router. Then attacker can now see all of the victim's traffic. If the victim is browsing sites that are not using proper encryption techniques, then the attacker is able to view all of the data, such as his or her login credentials to various sites.

This specific MitM attack mentioned is referred to as ARP poisoning. Address Resolution Protocol (ARP) is part of the link layer in the TCP/IP protocol suite. Its job is to resolve network layer addresses (IP addresses) into link layer addresses (MAC addresses). ARP works with request and reply messages. First, the computer wanting to send data sends an ARP request, which is essentially asking what computer is associated with the

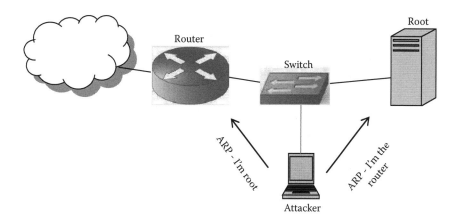

FIGURE 16.3 Man-in-the-middle attacks show how all user and network traffic can pass through an attacker's device.

destination IP address. The receiving computer then sends an ARP reply, letting the sending computer know that it has that IP address, and tells the sending computer its MAC address.

The ARP poisoning attack works by the attacker tricking the victim's computer into thinking that the attacker's computer is the router. This is done by having the attacker's computer send an ARP reply to the victim's computer with the router's IP address, but the attacker's MAC address. Now the victim thinks that all the traffic it needs to send to the router, which is then most likely being sent on to the Internet, should be sent to the attacker's computer. Essentially the same thing is done between the attacker's computer and the router—tricking the router into thinking that the attacker's computer is the victim's computer, thus sending any receiving data from the router, or the Internet, first to the attacker's computer, which then forwards it onto the victim's computer. As far as the victim is concerned, everything appears as it should because he or she is still able to send and receive data just fine.

There is one caveat with this attack. What if the victim is sending encrypted data? How is the attacker able to read the encrypted data? There is a way around. Instead of the attacker trying to read the encrypted data, it can make sure the victim isn't using a secure connection, which encrypts the data. If the victim is using the HTTPS (vs. HTTP) protocol to access sites, then the data are encrypted. The S in HTTPS stands for Secure. HTTPS uses Secure Socket Layer (SSL) or Transport Layer Security (TLS), cryptographic protocols, to encrypt the data.

These protocols are used to encrypt web traffic as it passes between the client and its web server destination. As mentioned, since the data are encrypted, the attacker is unable to read it. There are ways to strip these protocols so that the victim uses HTTP instead of HTTPS, making the traffic unencrypted and viewable by the attacker. For this type of attack (in combination with a MitM attack) to work, however, the website must have an HTTP version as well. Secure sites such as Twitter.com (and therefore Twitter mobile apps) do not contain an HTTP version of their site. This forces users to use the more secure HTTPS version. This attack works by essentially redirecting the victim to the HTTP version of the site, thus making him or her send unencrypted traffic through the attacker's computer. This is done by switching which port the victim is using. HTTP traffic uses port 80, whereas HTTPS traffic uses port 443. So when the victim enters his or her login credentials to a website, it will switch the port that data are sent from to port 80 so that the data are not encrypted and can be easily sniffed by the attacker. Thus, the attacker now knows the victim's login credentials for that website. For how to implement these types of attacks, be sure to check out tools like Easy-Creds, Ettercap, and SSLstrip, which are all free.

16.6.3 Mobile Monitoring

With any form of communication, there always has been and always will be one common interest: privacy. Whom do I want to be able to see what I'm sending, or more importantly, whom do I *not want* to see what I am sending? How do I keep all my personal data private? As mobile devices are becoming more and more popular, not to mention more capable, privacy has become an increasing concern. With this, the monitoring of mobile devices has become a matter of great interest. There has been rising news coverage in the last couple of years discussing companies' ability to hack in to mobile device voicemails, as well as government wiretapping of devices.

The government has many different ways of accessing mobile devices. One method is known as a roving bug. This method involves activating a cell phone from a remote location and turning its microphone into a listening device. With almost all phones nowadays having built-in GPS, this makes it easy for the government to track mobile devices, and therefore their owners. Take a look at the apps used on your phone. If you ignore the warning or messages that the app displays, you might be surprised how many different apps are tracking your location. With all these different applications tracking your location, countless databases exist with stored information pertaining to your location.

It's not just voicemails, texts, and locations though. It's the data as well. With the recent release of classified documents from ex-employee of the NSA Edward Snowden, it has been revealed how the NSA has infiltrated a lot of the major data communication companies like Google. One thing to keep in mind is that all data are being funneled through your mobile carrier company. Once it has them, it pretty much knows everything about you—whom you call and text the most often, what sites you frequently visit, and physically where you have been. So how do you prevent anyone from tapping your mobile device? The only true defense is to remove the battery. If the device does not have a power source, no action can be performed.

CHAPTER 17

Corporate Security for Mobile Devices

Lauren Collins

CONTENTS

S MARTPHONES AND TABLETS are used for business and personal use and are growing in popularity. The increase in popularity shows a shift in the preference of a tablet as the chosen device for work due to its high functionality. The trend of employees selecting tablets as their primary work device is expected to continue at a rate of 5% over the next year. Organizations that have workers only connecting over wireless expect more than 10% of employees to fall into this category over the next year, as shown in Figure 17.1. This trend will grow as time goes on, forcing companies to build out wireless local area network (WLAN) as their primary means of access.

Productivity increases when the proper tools, devices, and technology are available to an employee. A majority of companies not only encourage the use of mobile devices, but also allow employees to use their personal mobile devices to access corporate data on and off the corporate network. Corporations fail to recognize that mobile devices represent a threat to data security. Users give greater precedence to their own rights on a device than to the employer's need to protect data. Unfortunately, many businesses lack visibility into the types of devices their employees are using, as well as the type of data accessed on those devices.

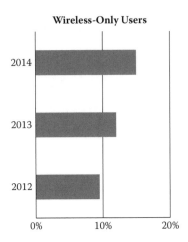

Wireless-Only Users

FIGURE 17.1 Wireless-only users are on the rise by roughly 5% year by year.

The innovation of mobile devices presents security implications that are hardly visible to C-level executives and the IT department. Employees expect to have access to corporate resources on their mobile devices everywhere and all the time, adding increased support concerns for the IT department. While endorsing this type of access may align with the business's strategy and enhance employee productivity, it also creates security concerns.

After these strategies have been defined, mobile application requirements should be considered. If the corporation is deploying an in-house app, ensure a mobile application management (MAM) system is in place. As an alternative to a MAM, security or network administrators can compare the WLAN access list against the corporate directory. Reviewing these lists regularly provides criteria for known employee devices compared to the number of devices actually connected. Without monitoring mobile devices connected to the network, there is no way of ensuring compliance or security. It is essential to ensure the proper employees have the proper access delegated to them, whether based off their job role, function, or group. To further compound the issue, mobile devices can connect to the public Internet and local network directly using separate mediums. For example, a mobile device may be associated with the local wireless network belonging to the corporation and a 3G/4G/LTE connection to the Internet.

Junior members of an IT department view security and policies as a barrier that gets in the way of productivity and efficiency. This attitude directly impacts and negatively affects how well a corporation's network and data are secured or how great the potential is for penetration. The attitudes and experiences of IT are contagious, and a negative perception can spread to other business units. Most members of IT work to connect users to resources, and in this modern era, more and more data must be quickly and easily accessible. A common goal across IT departments is to provide ease of access to information and to make the lives of employees easier. Consider the role of a security administrator, which is very different from other positions throughout the company. Employees are simply attempting to access resources to do their jobs, and all too often, IT will implement an untested change or push out a spontaneous policy. After such changes are made, there is a swarm of helpdesk tickets and complaints that arise (e.g., "Yesterday I could perform this task, and today I cannot"). End users have no idea whether a change is good or bad; they only know it may present a new, unnecessary obstacle to their job. From their perspective, security is a bad idea, and is someone else's decision to impede their workflow.

Educating all employees, particularly those in IT, is essential so that they understand the importance of security. The decisions made on a daily basis directly impact and affect corporate network and data security. While most employees are not malicious in their activities and actions, the majority lack security education. IT department decisions have as direct an effect on successful implementations as does cultivating user awareness of a new security practice. Positive attitudes and a direct approach in concurrence with being vigilant in the enforcement of new policies will keep security at the forefront of employees' daily activities. Time spent educating IT staff and other business units will ensure they are security literate and understand the logic behind implementing policies and best practices in accordance with their employer's procedures.

17.1 DEFINING A POLICY

The simplest policies to implement give full visibility and control over all devices and users who connect to the company network and account for company devices connected to other networks. Legacy policies support control and blocking of content, which restrict the employees and users from accessing predefined data and prevent security breaches. Today there is a need for a risk management policy definition. Once authenticated and connected, mobile devices present the risk of data leakage and compromise security models that are defined to protect both sensitive and proprietary data.

Most corporations require employees to sign an acceptable use policy on their first day of employment.

Mobile Device and Usage Policy

I. Introduction

Mobile devices, such as smartphones and tablets, are important tools for the organization, and they are used to support and achieve business goals.

However, mobile devices also represent a significant risk to information and data security. If the appropriate security applications and procedures are not applied, they can be a conduit for unauthorized access to the organization's data and IT infrastructure. Subsequently, this can lead to data leakage and system infection.

[Company X] has a requirement to protect its information assets in order to safeguard its customers, intellectual property, and reputation. This document outlines a set of practices and requirements for the safe use of mobile devices.

II. Scope

a. This includes smartphones, tablets, and all mobile devices, whether owned by [Company X] or owned by its employees with access to corporate networks, data, and systems, not including corporate IT-managed laptops.

b. Exemptions: Where there is a business need to be exempt from this policy, whether it is too costly, too complex, or adversely impacting other business requirements, security management must authorize and conduct a risk assessment.

III. Policy

Technical Requirements

a. Devices must use the following operating systems: Android™ Jelly Bean or later, iOS or above. [List all device versions relevant to the organization.]

b. Devices must store all user-saved passwords in an encrypted password store.

c. Devices must be configured with a password that complies with [Company X's] password policy. This password must not be the same as any other credentials used within the organization.

d. With the exception of those devices managed by IT, devices are not allowed to be connected directly to the internal corporate network.

User Requirements

a. Users must only load data essential to their role onto their mobile device(s).

b. Users must report all lost or stolen devices to [Company X] IT immediately.

c. If a user suspects that unauthorized access to company data has taken place via a mobile device, the user must report the incident in alignment with [Company X's] incident handling process.

d. Devices must not be jailbroken* or have any software/firmware installed that is designed to gain access to functionality not intended to be exposed to the user.

e. Users must not load pirated software or illegal content onto their devices.

f. Applications must only be installed from official platform owner-approved sources. Installation of code from untrusted sources is forbidden. If you are unsure if an application is from an approved source, contact [Company X] IT.

g. Devices must be updated with manufacturer- or network-provided patches. At a minimum, patches should be checked for weekly and applied at least once a month.

h. Devices must not be connected to a PC that does not have updated and enabled antivirus and malware protection, or a PC that does not comply with corporate policy.

i. Devices must be encrypted in line with [Company X's] compliance standards.

j. Users must be cautious when merging personal and work email accounts on their devices. They must take particular care to ensure that company data are only sent through the corporate email system. If a user suspects that company

---

* To jailbreak a mobile device is to remove the limitations imposed by the manufacturer. This gives access to the operating system, thus unlocking all features and facilitating the installation of unauthorized software.

data have been sent from a personal email account, either in body text or as an attachment, he or she must notify [Company X] IT immediately.

k. Users must not use corporate workstations to back up or synchronize device content, such as media files, unless such content is required for legitimate business purposes.

Note: The sample policy was provided by the author solely as a demonstration of what a corporate policy may look like. The authors are not attorneys, and this document should not be construed in any way, shape, or form, be it verbal, written, or digital, to represent legal advice. The authors do not warrant that this sample document will provide full or complete coverage that a company may require, as every corporation's needs differ.

The policy outlines the corporation's stance on data governance, the purpose of which is to protect proprietary data and ensure the corporate network is not compromised due to the use of personal devices or personal use of company-provided devices. Still, corporations that do not recognize the threat these mobile devices represent to data security face challenges, mainly users who allocate greater precedence to personal use of their devices rather than the employer's need to protect data. By assessing employee activities and usage, IT departments and security administrators are in a better position to consider mobile device security strategies, without reducing functionality or productivity.

17.2 COMPLIANCE

Extending convenience to employees weakens the virtual walls of network security. Companies may also have several employees who are inundated with social media, dabble in freeware, or have steadfast access to other types of unstructured data. Also consider the management and monitoring of mobile devices, and how they overlap with compliance initiatives, such as the Health Insurance Portability and Accountability Act (HIPAA) or payment card industry (PCI). If the corporation were to perform data analytics on all activity taking place on employee devices, it may ask for passwords to their social media or accounts. Since no privacy laws have yet been instituted for something of this stature, it could be an uncomfortable conversation.

Proactively implementing security controls not only is leveraged internally, but also can suffice the ongoing needs of compliance. For example, a trading firm is located in the United States and trades multiple foreign exchanges. These foreign exchanges deem it mandatory to obtain a local bank account, making it an offshore account for the trading firm. If this trading firm has clients, it makes sense to monitor activity and communicate potential threats and breaches. Client A poses the question: "How do I know my money will be safe moving from the United States to Brazil?" Since the compliance team was aware client A traded both U.S. and Brazilian exchanges at the time they signed up with the firm, this pertinent information could be furnished up front rather than only at the time of client A's concern.

Being proactive proves successful to your internal business units as well as external customers with the knowledge that complicated issues are being addressed and accounted for. Whether the corporation has only internal stakeholders or both internal and external stakeholders, the last appearance a CSO (chief security officer) wants an invitation to is a meeting with government or large clients presenting security initiatives at *their* board meetings. This type of invitation usually comes about in the reactionary mode, and has the team scrambling for answers to get a plan together quickly. Any corporation that has defined its security policy, and effectively maintains that policy, possesses ample data and metrics available. Officers of the corporation are not likely to ask for percentages of failed patching efforts, but they are interested in knowing *why* an issue is occurring. Again, what are the business reasons driving the processes surrounding the organization as a whole, and then its business units? Are there sacrifices being made in order for the business to run efficiently? If so, a risk-based security approach would be the best way to go. A common performance measurement displayed across organization and business units is key performance indicators (KPIs). KPIs show the success of a particular activity or project; the success is completely dependent upon a repeatable process, utilizing the periodic achievement of a goal. The evaluation of KPIs allows for a definitive set of values against which to measure other quantitative indicators. Where security and compliance are concerned, key risk indicators (KRIs) are useful to indicate the risk of an activity. KRIs provide a forewarning, or probability of potential events that may lead to an adverse impact, rather than the intended positive result of a goal. Trending and visual analysis utilizing KPIs and KRIs allow a corporation to obtain a competitive advantage and get ahead of issues before they fully develop.

This chapter presented a sample policy a corporation may furnish to employees. Compliance differs from firm to firm, based off their business model, whether they are public or privately held, and the manner in which business ensues.

When determining policy compliance, there are several factors to consider:

- Type of phones that are offered and supported:

  - BlackBerry® Enterprise Server (BES) is 100% compliant.

  - iPhone® and Windows® Mobile are only 100% compliant using ActiveSync.

  - Non-Blackberry devices can utilize third-party solutions— Good, Mobile Iron.

- Individual liability vs. corporate liability: It is the firm's policy that the wireless numbers associated to all firm-issued mobile devices are owned by the firm. There will not be approval to release an employee's number upon departure from the firm. Employees may opt to transfer their personal number to a firm-issued mobile device. Once an employee separates from the company and requests to have the number transferred back, approval will not be granted.

- Operating vehicles: Employees are prohibited from using firm-issued mobile devices while driving. Prior to using a mobile device, employees are expected to abide by applicable laws, regardless of whether the vehicle is owned by the firm.

- Camera functionality: Capturing and storing inappropriate photographs or images is forbidden pursuant to the firm's use of Internet and email policy. Employees are responsible for familiarizing themselves with said policies, and also for abiding by any specific business unit controls that may apply.

17.3 MANAGING DEVICES

Unlike desktops and laptops, the operating system on tablets does not have Windows. Therefore, the infrastructure and security constraints differ greatly between Windows and iOS, Google®, or Android™ devices. When giving the user an option for a mobile device, employees' preferences are either iOS, Google, or Android. These options are phasing out other platforms such as Blackberry. Still, corporations are using ActiveSync,

Blackberry Enterprise Server, or Google Sync to enforce security policies on mobile devices, such as requiring PINs, passwords, and other levels of authentication and credentials for exchange access. However, ActiveSync and BlackBerry Enterprise Server are not able to manage mobile devices or any of the applications running on smartphones and tablets. By first defining what needs to be managed, and then creating and implementing a policy, it is more feasible to obtain a practical third-party provider to manage mobile devices.

Mobile device management (MDM) enables network security professionals to monitor and manage devices across an enterprise network. Security professionals and IT managers have a variety of options and vendors to select from with MDM, as device management solutions are evolving at the pace of consumers' and businesses' purchase trends, as shown in Figure 17.2.

The evolution of mobile devices has prompted vendors to offer a wide range of options to meet the needs of device management and support the mix of technologies:

- Native OS administration

- On-phone hypervisors

- Network-centric controls

- Secure containers

- Virtual desktop infrastructure (VDI)

Figure 17.3 reflects the business needs for MDM and MAM. Although corporations differ, the majority are implementing MDM and MAM based off app management, where asset tracking used to rank the highest.

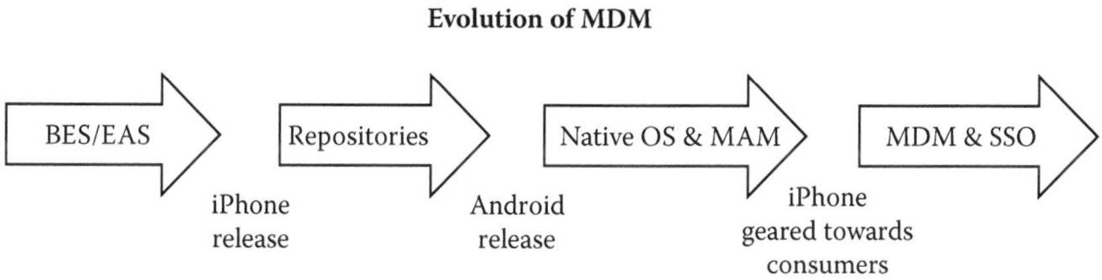

Evolution of MDM

FIGURE 17.2 Corporations must account for platform releases and shift their management tools to keep up with technology.

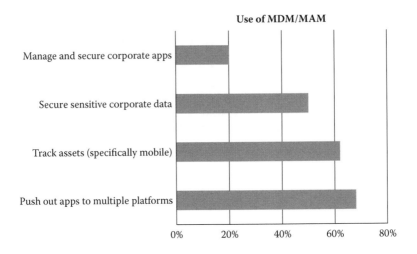

FIGURE 17.3 Corporations are shifting the purpose for implementation of MDM/MAM. App management accounts for 67%.

As Table 17.1 illustrates, assorted technologies address explicit security requirements, while accomplishing business goals. Earlier revisions of MDM solutions were intended to support iPhone and Android devices, utilizing secure repository methodology. Briefly, these solutions provided a secure, managed template for users to log in and access email, proprietary tools, and company data. When security supersedes the appeal of a device and its functionality, secure repositories interfere with user experience (UX). Then, mobile devices did not support single sign-on (SSO), which imposes users to enter their credentials every time their device wakes from sleeping. Today, offering functionality and security revolves around and reflects the growing importance of apps in the workplace.

While UX and security have improved, what about those personal devices IT is unaware of and has not accounted for in the security policies? For those devices that are not provisioned for work, they do not have a client installed, or the client is unknown to IT. Consequently, many software MDM vendors partner with WLAN vendors to integrate network access controls (NACs). Chief information security officers (CISOs) may elect to reevaluate the infrastructure and policy requirements of providing home office and field access to their employees.

For instance, Figure 17.4 illustrates the architecture of the Cisco® Identity Services Engine (ISE) integration. ISE has the capacity to integrate with third-party MDM tools, such as Good Technology and Mobile Iron. Since there are so many vendors to choose from, it is imperative that IT selects

TABLE 17.1   Feature Set Breakdown to Determine Benefits, Challenges, and Business Case for Deploying MDM to Manage Mobile Devices in a Corporation

	Benefits	Challenges	Business Case
Secure repository	• Protects company data • Creates consistent UX/UI across all devices • Data are ephemeral on device or locked in repository • Protected repository may supersede OS functionality	• OS can break when upgraded, alters capability • May have proprietary browser and document editing apps • Can break device UI/UX • Doesn't leverage OS native MDM capabilities	• Manage documents and apps within the limitations of repository
Native-centric controls	• OS supported by default • Sustain UX/UI • Features and capabilities tested by OS originators • Separation of enterprise and personal email	• Functionality differs across device platforms • OS can become unstable once fragmentation occurs • Some platforms are not compliant	• OS design can be an advantage if there's a specific feature set a particular platform offers
Network-based MDM	• Device is managed by the network team; user cannot delete or make changes • All devices are supported • Granularity to manage data, access • Easily upgradeable and accounts for the future	• Unable to initiate a device wipe • Unable to control app circulation • Only works within the corporate network • Particular functionality requires additional vendor components: controllers/access points	• Supports bring your own device (BYOD) and independent platform management
Virtual machines	• One device supports multiple users • Data are ephemeral on device or locked in repository • Total separation of identity, apps, and data • Augment existing desktop and apps to mobile devices	• Two devices per user (personal and enterprise) • Resource-intensive • Shorter battery life • Touch-based devices not ideal for content design	• Centralized control over virtual enabled content

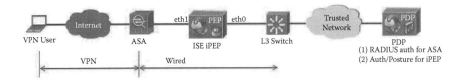

FIGURE 17.4 An ISE deployment allows an administrator to make proactive governance decisions by correlating identity to various network devices.

a solution that will allow for self-service enrollment of mobile devices, which increases efficiency and accuracy across the entire corporation.

A predefined pool of apps can be selected and allocated by management, and that pool has the ability to share data with each other securely. The templates that are applied allow for the transparency to manage the device from the status of a corporate directory, which may be based on an employee's status. As long as the device is connected to a network, not necessarily the corporate network, devices may be remotely wiped or upgraded, and even have the capability to accept a new policy that was pushed out unbeknownst to the user.

17.4 LOST AND STOLEN DEVICES

Lost and stolen devices are among the many new challenges of securing mobile devices. As discussed throughout this book, third-party tools and solutions allow for mobile device management (MDM) and mobile application management (MAM). Management of mobile devices allows IT administrators to define mobile device policies and provision software (apps).

Relevant to a lost or stolen device, MDM performs and facilitates the following features:

- Remote lock and wipe of device
  - Enterprise data definitions allowing for selective wipe and privacy policies
- Lockdown security for the device
- Mobile device location and recovery
- Enforce digital encryption policy for the mobile device and media components, such as secure digital (SD) cards
- Access control, device visibility, and blocking of email retrieval

- Digital certificate distribution

- App accounting, distribution, blacklist

- Password enforcement

- Secure administration with role-based access, group-based actions, persistent logging, and audit trails

Referencing Figures 17.5 and 17.6, lower-level apps exist and provide similar features as listed above, such as Find My iPhone or Where's My Droid, that allow a user to locate the mobile device on a map, display a message on the device, play a sound, and remotely lock and wipe the device.

At some point in the life cycle of a mobile device or an employee's duration of employment with the corporation, it becomes necessary to terminate access to the device due to a lost or stolen device, employee termination, or other modifications. Network and security administrators must have the ability to rapidly revoke access to any mobile device and to remotely wipe some or all of the data and apps on the device.

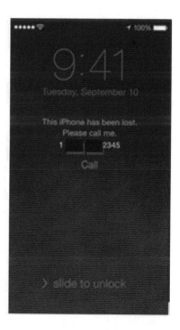

FIGURE 17.5 Mobile devices iPhone and iPad have a less sophisticated app to locate the device if lost or stolen.

FIGURE 17.6   Mobile devices with Android OS Where's My Droid to locate the device if lost or stolen.

CHAPTER 18

Mobile Device Management (MDM)

Lauren Collins

CONTENTS

FOR SMALLER CORPORATIONS AND FOR PERSONAL DEVICES, less sophisticated apps like Find My iPhone may be all that is necessary to locate and wipe a device. However, identity, authentication, and systemwide visibility disclose who is on a network and the type of device that is being used. These details are extremely important for large organizations that need to secure proprietary data and lock down devices. Mobile device management (MDM) incorporates greater visibility and control of the endpoint device. A mobile framework must manage not only the business applications, but also the devices themselves. Conducting a mobile assessment should be an IT department's first step.

Organizations should evaluate the following:

- How do you secure devices, data, and apps?

- What type of policies, governance, and support do organizations need when supplying new mobile device technology?

- What access and storage limitations must be implemented and enforced to protect proprietary information while enabling access to low-risk edge applications?

After conducting a comprehensive assessment, organizations can identify best use cases for mobile technology. The next step is to develop a long-term strategy for solving targeted challenges. This step-by-step approach allows an organization to roll out features based on high-priority needs, and avoid engrossing in unnecessary and time-consuming proofs of concept.

In Chapter 17, policies, protocols, and processes defined how mobile devices can be securely integrated into a corporate environment. For instance, when faced with the question of buying or owning the device, a company takes the hybrid approach. The organization will allow employees to buy and use the device of their choice, whether it is an iPhone®, Android™, or Blackberry®, or iPad® or other tablets. However, as organizations install enterprise mobile apps on those devices, they must also consider how, and to what extent, they will support the hardware that runs the software. The line between work and personal life continues to blur, and the users supported demand a consistent and personalized experience no matter what device they are using. MDM solutions can take a device-centric approach to management, while a solution like the Cisco® Identity Services Engine (ISE) takes a network-centric approach. Many administrators choose to employ both solutions in tandem to integrate services and functionality as part of the bring your own device (BYOD) solution.

18.1 MDM SELECTION

MDM management software delivers IT requests over the air to company-issued or employee-owned devices running everything from Apple® iOS to Google® Android. MDM products are rapidly evolving to keep up with the torrent of new devices evolving in the market, updated mobile operating systems, and increasingly complex business needs. With many diverse products to choose from, selecting the right software can be daunting. Start any MDM product evaluation by narrowing your choices to those capable of supporting mobile devices and operating systems of strategic importance to your workforce. Accept that MDM products may not support older devices, while some new devices may not yet be accommodated by MDM products. Rather than focusing on covering every device, pursue a solution that includes depth of available capabilities. Several MDM vendors provide legacy management for older devices; several MDM products present limited control over nearly every device using Exchange ActiveSync. By reviewing each product's track record for support of newly released devices and OS versions, past performance can indicate future expandability and anticipated time to market.

Now that you have compiled a list of products that are capable of managing the majority of your organization's devices, what are the capabilities offered for each mobile device OS? On paper, MDM products tend to be quite similar. For instance, each product on the list should offer device policy management. Any product missing this basic MDM capability should be excluded immediately. While this may seem self-explanatory, many offerings excel at only one entity, such as secure enterprise email or mobile expense management, misplacing the product on the MDM product lists.

In reality, the lack of industry standardization is a fundamental challenge. MDM vendors use buzzwords to compare similar features and functionality, which may impede decision making or sway an opinion at first glance when comparing products. Instead, list your workforce requirements and use them to compare what each MDM product offers for each mobile OS. Table 18.1 outlines the rudimentary resources any MDM product should offer, aligning common IT tasks with correlating features. Observe that supported tasks and features differ across various MDM products. This is where you will begin to appreciate each MDM product's fit for your workforce. For instance, all products will offer device enrollment. Historically, IT enrolled company-issued devices, individually or in volume. Currently, users can visit self-enrollment portals and enroll their device, supporting BYOD, and if approved, the device is automatically provisioned with device policies. Some prefer an enrollment portal that integrates with Active Directory, taking advantage of single sign-on. Rather than requiring IT to define the same management policies repeatedly for every user in a group, provision devices with group-based policies.

Evaluating the features related to each mobile OS is just as important as assessing the tasks and features of MDM products. All MDM products offer the ability to configure PINs and passwords; however, each mobile OS determines the PIN or password length, strength, and complexity. While a MDM tool may be able to deliver and apply universal tools across mobile devices, the products cannot mask the differences in mobile OSes. These

TABLE 18.1   Fundamental Mobile Device Management Features

Capability	Description	Tasks	Features
Device policy management	Obtain/establish device features and restrictions to assert and enforce IT-defined policies	• Define policies • Provision policies • Enforce policies • Maintain policies	• Acceptance criteria • Group/location policies • Policy refresh • Compliance checks • Enforcement actions
Security management	Protect and access the integrity of enrolled devices	• Configure controls • Enforce controls • Check integrity • Detect breaches	• PIN/password • Inactivity timeout • Login failure • Data encryption • Device restrictions • Secure Wi-Fi, VPN, email • Jailbreak detection • Blacklist enforcement
Inventory management	Create and maintain a database of enrolled devices and their properties	• Device enrollment • Asset tracking • Decommissioning	• Self-enrollment • Directory integration • Acceptable use policy • Asset details • Change history • Remote wipe • Backup/restore
Monitoring and reporting	Deliver real-time and historical visibility in enrolled devices and their activities	• Real-time stats • Alert notifications • Event logging • Device location	• Configurable dashboard • GPS tracking and mapping • Canned/custom reports • Summary and detail views

products can also notify you when particular rules are not supported on a given OS version. Certain MDM products can also automatically check devices and quarantine or unenroll those that do not comply with policies.

Carefully consider how criteria are set and enforced, and what degree of control and automation an MDM product delivers. If an employee installs a blacklisted app, a MDM product is capable of remotely wiping the device or simply sending a notification to the user that the app is banned and should be removed. The proper action may depend on the type of mobile device as well as the user. Comprehensive MDM products allow IT a scope of useful administrative actions, along with the capacity to apply policies wisely.

18.2 DEPLOYMENT MODELS

The mobile device management features illustrated throughout this chapter can be deployed in multiple ways, as shown in Figure 18.1. A conventional deployment model involves installing MDM software on site, on a dedicated server operated and owned by IT and located in a corporate data center or hosting facility. Multiple large organizations continue to run this model as it streamlines integration with other enterprise services such as directories, mail servers, and file servers. Cloud offerings have prompted growth of alternative models. Specifically, enterprises now consider deploying MDM software on private or public cloud servers, taking advantage of network redundancy, high availability, and limitless scalability. Most MDM products can be deployed in this fashion.

Software as a service (SaaS) is another prevalent model among small to medium-sized businesses. A SaaS model allows MDM vendors to install their software on their multitenant servers, selling MDM resources as public cloud services. SaaS can be an influential means for MDM evaluation. Many companies utilize this pay-as-you-go alternative when lightly managing a large number of devices under BYOD. A pilot of a MDM

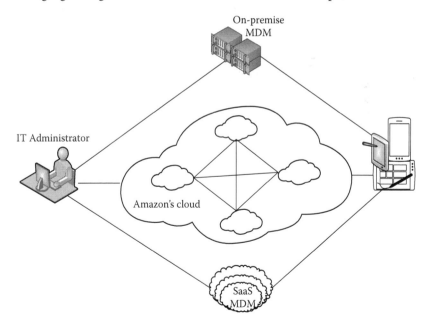

FIGURE 18.1 Corporate IT can deliver mobile device management by deploying software on the premises, deploying software in a cloud, or purchasing SaaS management.

product can be quickly set up in a SaaS environment. Then in the long run, if your organization opts to manage its own MDM server, the pilot can easily transition to production.

18.2.1 Device-Centric Approach

Users want to access corporate applications from anywhere, using any device they choose: laptop, smartphone, tablet, or PC. By delivering unified application and device management, both in the cloud and on the premises, organizations can support BYOD by implementing a single solution. Manufacturers, such as Apple and Microsoft, have a platform available that can automatically detect devices in the network, and send the device settings for immediate and continued visibility. This process is fully automated, keeps a history of devices, and can push updates as well as wipe a device.

To communicate with an iOS mobile device, all MDM servers use an Apple push notification service to silently prompt the device to check in for management. When a secure connection is established between the device and server, each management task is carried out by the integrated MDM framework in iOS. The framework enables MDM servers to contact devices simultaneously without impacting performance, battery life, or workload. Fortunately, there is no need for each MDM vendor to implement custom agents to perform management tasks. Figure 18.2 illustrates the process to enroll and manage an Apple mobile device utilizing a third-party solution. Multiple server platforms, management consoles, and workflow options are available.

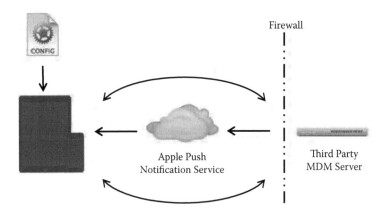

FIGURE 18.2   Apple's enrollment and management process for mobile devices.

Below is an overview of the enrollment and management process, shown in Figure 18.2:

1. A configuration profile containing MDM server specifications is sent to the mobile device.

2. The user installs the profile to allow the mobile device to be managed.

3. Device enrollment ensues as the profile is installed. The server validates the mobile device and grants access.

4. The MDM server sends a push notification, initiating the mobile device to check for tasks or other inquiries.

5. The iOS mobile device connects directly to the server over HTTPS, and the MDM server sends updates or requests information.

If an organization elects to implement many management servers, or only one for Apple devices, there is a Profile Manager available from Apple. Profile Manager is used widely by organizations that employ several Apple devices, and schools have recently been added to this list. An OS X Server runs Profile Manager, which maintains all the requirements of the organization and does not necessitate additional client licensing to maintain the OS X Server. Many IT administrators find Profile Manager the easiest to set up and the simplest to maintain and manage. There is also a self-service web portal where users can download and install new configuration profiles, clear passcodes, and remotely lock or wipe their mobile device.

18.2.2 Network-Centric Approach

As mentioned in Chapter 17, the Cisco Identity Services Engine (ISE) provides coherent enforcement for policies on wired and wireless corporate networks. Cisco ISE integrates authentication, authorization, and accounting, also referred to as AAA services, as well as profiling, posture, and guest services, to gather real-time information from networks, users, and devices.

The ISE deployment is based on node roles, and there are four node roles offered:

1. Admin: The Admin node allows for central management of the ISE deployment. The majority of configuration takes place within this node.

2. Monitoring: The Monitoring node collects all monitoring events relevant to events granting or denying authentication and authorization.

3. Policy: The Policy node communicates with the endpoints and has the ability to make decisions based on authentication and authorization. This is your radius server.

4. Inline Posture: The Inline role is used with devices that do not have support for change of authority (CoA). The radius CoA feature provides a function to alter features of authentication, authorization, and accounting (AAA) sessions after authentication. Most features of ISE require CoA. If a policy changes for a user or a group in AAA, an admin can send the radius CoA packets from the AAA server to authenticate again and invoke the new policy. The inline appliance is physical.

Depending on the ISE design, appliance roles can be used on a single appliance or they can be combined. The more policy nodes in your network, the larger the node and the more there is to administer. There is a maximum of two admin nodes and two monitoring nodes.

Here are key considerations for implementation of ISE:

- Licensing: See Figure 18.3.

- Design: Ensure the number of nodes is consistent with the license count for allowed endpoints.

  - Licensing is based on the concurrent number of endpoints.

  - An endpoint is considered an authenticated device: PC, tablet, phone, printer, iPad, etc.

  - Profiling (detection of device type) and posture (health of an endpoint) are only included in the advanced license.

ISE Wireless License	ISE Base License	ISE Advanced License
Base + advanced	Are my endpoints authorized?	Are my endpoints compliant?
• All base services • All advanced services	• Guest provisioning • Link encryption policies • Authentication and authorization	• Device profiting • Security group access • Endpoint on-boarding

FIGURE 18.3 Licensing is based on concurrent endpoints and also on features. The wireless license includes base and advanced features, but for wireless only.

- Determine the maximum number of endpoints with all roles.
    - One ISE appliance supports a maximum of 2000 endpoints.
    - Two ISE appliances support a maximum of 4000 endpoints.
- Performance and bandwidth requirements:
    - Check the vendor website for current deployment values.
- Limitations:
    - Cisco ISE hostnames are limited to 15 characters or less when using Microsoft Active Directory on the network.

CASE 18.1

A medium-sized organization has 500 employees. Each employee has two to four devices on the corporate network. The mobile devices are both corporate-issued and personally owned devices. The apps run on the mobile devices incorporate both proprietary apps and personal apps.

SOLUTION: CISCO ASA 5515X

The Cisco ASA 5515X, a hardware device, would easily cover minimal security and performance requirements, as well as allow for some growth for the future. The Cisco AnyConnect Essentials, a virtual private network (VPN) client/app, and a number of mobile licenses will allow iOS, Android™, and Windows® devices to connect securely. *Note*: There's an AnyConnect app to install, shown in Figure 18.4, that supports the latest IPsec version (IKEv2) and Secure Socket Layer (SSL) VPNs.

The Cisco AnyConnect Secure Mobility Solution offers a thorough, highly secure enterprise mobility solution. When implementing this security solution to manage devices, users are allowed to access the network, and easily and securely use applications and information pertinent to their jobs using a device of their choice, whether it is a laptop, tablet, or phone (Figure 18.5).

The Cisco AnyConnect Secure Mobility Solution includes

- Logical connectivity that is always on
- Secure mobile access across the increasing number of managed and unmanaged mobile devices
- Security policy administration that is context aware, comprehensive, and preemptive

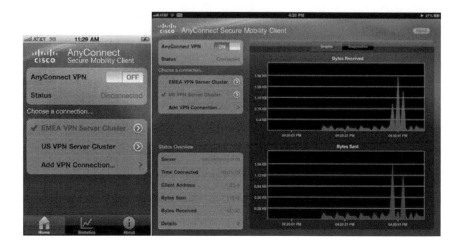

FIGURE 18.4   Cisco AnyConnect app provides access to business-critical applications.

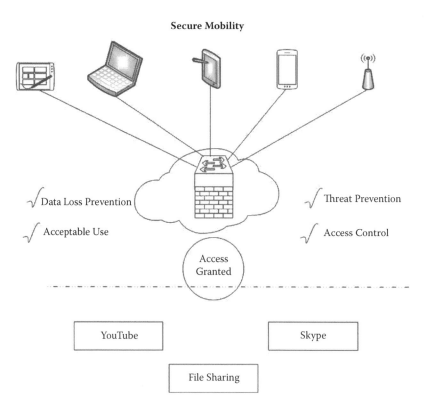

FIGURE 18.5   Access data on your device managed by your network.

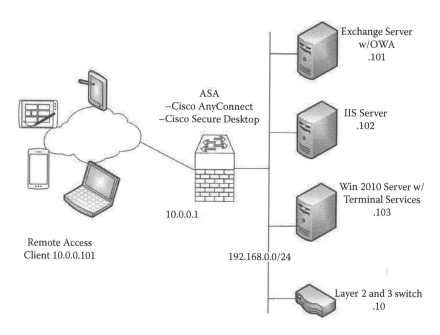

FIGURE 18.6 Scans are done on the devices with premium and advanced licensing.

OPTIONAL ADD-ON

For additional mobile and bring your own device (BYOD) function-ality, while implementing only the ASA, AnyConnect Premium and Advanced Endpoint Assessment licensing can be added. This option allows the AnyConnect VPN client software to do device scans for installed apps, registry entries, etc. (e.g., Is this really our lap-top? Are the AV signatures up to date? etc.). Premium licensing also enables the option of clientless SSL VPN. For example, a corporation could set up a clientless SSL VPN for accessing an internal web app from the Internet without any VPN software (Figure 18.6).

CASE 18.2

A large organization has 5000 employees. Each employee has one to four devices on the corporate network. The mobile devices are both corporate-issued and personally owned devices. The apps run on the mobile devices incorporate both proprietary apps and personal apps.

For more long-term granular security needs, Cisco has an add-on to use along with the ASA, Identity Services Engine (ISE). ISE can

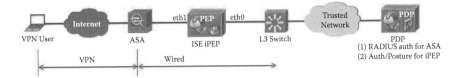

FIGURE 18.7 An ISE deployment allows an administrator to make proactive governance decisions by correlating identity to various network devices.

handle authentication and in-depth profiling and posture, and dynamic authorization for all the VPN clients, wired and wireless clients. This is Cisco's unified security solution. Figure 18.7 depicts what the setup would look like for ISE working with an ASA for VPN traffic.

The ISE design and pricing can be extremely complex, and can be implemented as a separate phase once an organization outgrows AAA. Some organizations decide on a 90-day trial of a Cisco ISE evaluation license to test and uncover ISE's potential. Large organizations with a need for granularity are already clear on their need for a solution like ISE, and would harvest all the benefits of ISE. No matter the size of an organization, there are consulting firms who are best suited to design and implement ISE based on a firm's needs and business objectives.

18.3 THIRD-PARTY SOLUTIONS

Numerous mobile IT vendors claim their solution is universal; the result obscures what really matters to an organization. When searching for and choosing a mobile IT vendor, here are a few guidelines to follow:

- Comprehensive solution across mobile devices, apps, and content
- Proven customer success
- Loyal and robust partner ecosystem

After considering the device and network-centric approaches, imagine applying this to each manufacturer and business need. Overall, a third-party solution may be the most comprehensive, best architected, and highly innovative option available.

18.3.1 Comprehensive Mobile Infrastructure

Rather than an organization depending on its IT department to research, architect, implement, and support a long-term platform to manage mobile

Identify SOC Report's Role and Purpose		
Report used by customers as a part of their compliance with Sarbanes–Oxley Act or similar law regulation	Yes	SOC 1 Report
Report used by customers and their auditors to plan and perform audits of customer's financial statements	Yes	SOC 1 Report
Report used by customers/stakeholders to gain confidence, place trust in service organization's system	Yes	SOC 2/3 Report
Make the report generally available or seal	Yes	SOC 3 Report
Do your customers have the need for and the ability to understand the details of the processing and controls at a service organization, the tests performed by the service auditor, and results of those tests?	Yes	SOC 2 Report
	No	SOC 3 Report

FIGURE 18.8   SOC role and purposes.

devices, sometimes it is better left to a third party. An organization may have a wide variety of devices: Apple, Windows, Blackberry, Android, and Google. The types of companies marketing third-party mobile management adhere to passing security audits for Service Organization Controls (SOC) 1, 2, and 3 Compliance Reports. Figure 18.8 provides some basic guidelines portraying each SOC role and purpose. SOC engagements report on controls at a service organization relevant to security, availability, integrity, confidentiality, and privacy. To obtain a SOC report, one or more of the following key system attributes must be addressed:

- **Security:** System is protected against both physical and logical unauthorized access.

- **Availability:** System is available for operation and use as pledged or established.

- **Confidentiality:** Information designated as confidential and private. Personal information may be collected, used, retained, disclosed, and disposed of in conformity with the commitments in the entity's privacy notice, and with criteria set forth in generally accepted privacy principles (GAPPs).

- **Processing integrity:** System processing is complete, accurate, timely, and authorized.

In addition to security and confidentiality, and a wide range of mobile device types and operating systems, there are many email solutions and apps to consider in the grand scheme of the mobile management solution. And in terms of back-end infrastructure, third-party solutions can integrate with an organization's existing solutions for email, content and identity management, and network access, ensuring that current processes and permissions adapt easily to the mobile environment.

18.3.2 Customer Success

Selecting the optimal mobile vendor is not just about technology, security, and compliance. It is also about proven experience in making customers successful. Mobile devices bring potential to increase workforce productivity, while also introducing substantial challenges for IT. As previously mentioned throughout this book, mobile technology means adapting to the new relationship between IT and the end user. IT can benefit greatly from third-party experience and its customers' success. IT has the increasing, continuing requirement to provide stable, secure services, but now in an environment with fewer boundaries and less direct IT control.

Businesses of all shapes, sizes, and locations trust third-party mobile management vendors to manage their devices and applications. Whether it is a pharmaceutical company, a global law firm, a large network of hospitals, or a technology company, all these companies are already realizing the benefits of mobility and are well on their way to becoming mobile-driven enterprises.

18.3.3 Partner Ecosystem

The most successful solutions and deployments are those that are built around supporting each unique, exclusive environment, not the other way around. Oftentimes a vendor will solicit solutions and work to fit those into an organization's culture, sometimes sacrificing obtuse pieces that do not fit the mold of a product. To deliver an innovative mobile device management solution, multiple solution providers collaborate to deliver a product that fits all aspects of a thriving enterprise. This type of ecosystem allows the support of an extensive range of apps, operating systems, devices, and deployment configurations. The options results represent the conservation of existing IT investments while bringing mobility into your current environment.

FIGURE 18.9 Third-party vendors enable end users the access they need and enterprise IT to support the environment.

Figure 18.9 illustrates a comprehensive model that third-party mobile device management can offer. Integrated provisioning and ongoing configuration imply the need for vigorous tools, such as scripts and self-provisioning applications, to automate network operations. There are third-party vendors who can automate and standardize business processes while consolidating IT tasks across global enterprises. Also, enterprises may create rich applications that leverage mobility and provide engaging experience to transform business processes. Turning to a third-party solution, developers can focus on the business value of their applications rather than focusing on how to secure and report on those apps. Companies like Cisco and Juniper offer protection from unknown, unauthorized devices. If IT needs a more granular approach to its policy management, access control, and monitoring, many vendors offer monitoring and reporting across all network tiers and endpoints. So much time and effort goes in to developing a security policy for an organization. Mobile devices throw an entirely new set of security standards and considerations into the equation. The need for business intelligence and streamlined network analytics becomes increasingly important as more mobile devices are introduced into an organization. By using more business intelligence, IT can analyze the entire strategy of mobile devices and adjust plans to fit the needs and usage patterns of employees. There are also options to align requirements for user authentication on mobile operating systems to provide single sign-on (SSO) capabilities that not only meet security demands, but also improve upon the user experience.

Take advantage of all opportunities to test capabilities and features, refine MDM policies, and welcome feedback from business units and participating employees on IT-defined requirements and how well any MDM product meets those needs. The evaluation period can assess critical product attributes such as usability, scalability, reliability, and support from IT and the MDM vendor.

Index

Printed and bound by CPI Group (UK) Ltd, Croydon, CR0 4YY

24/10/2024

01778283-0009